A HISTORY OF
POLAR
EXPLORATION
IN 50 OBJECTS

A HISTORY OF POLAR EXPLORATION IN 50 OBJECTS

FROM COOK'S CIRCUMNAVIGATIONS
TO THE AVIATION AGE

ANNE STRATHIE

This book is for, and in memory of, family members, friends, travelling companions and readers who have explored, recently or in the past, this wonderful yet fragile world we live in.

Inside front: Map from *Theatrum Orbis Terrarum*, by Abraham Ortelius, 1570; iStockphoto.
Inside back: Septentrionalium Terrarum descriptio, by Gerard Mercator, 1623; Wikimedia Commons.

First published 2024

The History Press
97 St George's Place, Cheltenham,
Gloucestershire, GL50 3QB
www.thehistorypress.co.uk

© Anne Strathie, 2024

The right of Anne Strathie to be identified as the Author of this work has been asserted in accordance with the Copyright, Designs and Patents Act 1988.

All rights reserved. No part of this book may be reprinted or reproduced or utilised in any form or by any electronic, mechanical or other means, now known or hereafter invented, including photocopying and recording, or in any information storage or retrieval system, without the permission in writing from the Publishers.

British Library Cataloguing in Publication Data.
A catalogue record for this book is available from the British Library.

ISBN 978 1 80399 105 4

Typesetting and origination by The History Press
Printed in Turkey by IMAK

Contents

Introduction   9

Part I: Laying Foundations   11
1   HMS *Resolution* in Pack Ice   13
2   Elizabeth Cook's Ditty Box   18
3   Scoresby's Barrel Crow's Nest   23
4   A Panorama of Spitsbergen   27
5   William Scoresby's Manuscript   32

Part II: Exploring North and South   37
6   Weddell's 'Sea Leopard of South Orkneys'   39
7   Edward Parry's Deck Watch   43
8   A Canister of Meat   47
9   James Ross's Career-Defining Portrait   52
10   Rossbank Magnetic Observatory, Hobart   57
11   A Great Icy Barrier   61
12   Francis Crozier's Penguin   65

Part III: The Northwest Passage: The Search Continues   69
13   A Daguerreotype   70
14   A Rock at Port Leopold   74
15   John Rae's Octant   79
16   A Graveyard on Beechey Island   84
17   Eleanor Gell's 'Franklin Search' Collection   88

Part IV: A New Start   95
18   A Photograph of Antarctic Icebergs   96
19   A Menu for a Banquet   100

20	Cornelius Hulott's *Resolute* Box	104
21	'On Board *Eira*': From the '*Eira* 1880' Album	110
22	*Illustrated London News* Front Page	116

Part V: Antarctica Revealed — 121
23	A Stereoview of Adrien de Gerlache and a Weddell Seal	122
24	Louis Bernacchi's Cape Adare Home	126
25	RRS *Discovery*	131
26	Edward Wilson's Portable Paintbox	136
27	A Postcard of Three Scottish Scientists	141
28	ARA *Uruguay*	146

Part VI: Striving for Polar Firsts — 151
29	Amundsen's Dip Circle	152
30	Shackleton's Sledging Compass	157
31	Deception Island	162
32	Matthew Henson's Fur Suit	167

Part VII: Southward Ho! — 173
33	Ponting's Kinematograph	174
34	A Samurai Sword	179
35	Mawson's Anemograph	183
36	Joseph Kinsey's Visitors' Book, April 1912	187
37	'Three Polar Stars' Photograph, January 1913	191
38	Henry 'Birdie' Bowers's Sledge Flag	195

Part VIII: 'White Warfare' and Testing Times — 201
39	An Expedition Prospectus	202
40	A Statue of Cheltenham's Local Hero	207
41	A Rock from Elephant Island	212

Part IX: The Age of Aviation — 217
42	An Avro Antarctic Baby	218
43	A Tribute to Shackleton from a 'Fan'	223
44	'Uranienborg': An Explorer's Refuge	228
45	Mawson's Gipsy Moth	234
46	A Young Explorer's Special Medal	239

Part X: Learning from the Past and Looking to the Future		245
47	A 'Polar Centre': The Scott Polar Research Institute Building	246
48	The *Erebus* Bell	251
49	An Expedition Hut	255
50	A Well-Travelled Crow's Nest	259

Conclusion	263
Appendix A: Terminology, etc.	265
Appendix B: Summary Timeline	267
Appendix C: Maps	270
Acknowledgements	280
Notes	283
Bibliography	303
Index	313

Introduction

A History of Polar Exploration in 50 Objects: From Cook's Circumnavigations to the Aviation Age is, like all histories, *a* history, rather than *the* history of its subject. The timespan of this book runs from the late 1860s to the early 1930s – a relatively short period, but one during which understanding of the polar regions increased rapidly. This era included, to paraphrase Charles Dickens (who has a walk-on part in this book), 'the best of times and the worst of times'. Although we no longer adhere to some of the values of those times, we still owe a debt of gratitude to those who travelled into the unknown in hopes of increasing the sum of human knowledge.

Some of the fifty selected objects were, as might be expected, used during polar expeditions. Others, by contrast, shed light on explorers' personal lives or ways in which they communicated their plans and findings to others, including the general public. Some objects are associated with world-famous explorers, others with those whose names are less widely known but who played their part in polar exploration history. The sizes of objects range from that of a tiny box to a massive ice-shelf. Some objects are highly crafted; others are more or less as found in remote polar regions. Some remain in their original locations; most can be viewed in person or online.

Authors are regularly asked why they decided to write a particular book. This book emerged from my increasing realisation that many of the Antarctic explorers, scientists and photographers who featured in my previous books were also interested in or explored Arctic regions.

As to the format of the book, I have long admired and learned much from Neil MacGregor's *A History of the World in 100 Objects* and felt that a book using his template suited this topic and might appeal to both new readers and those already interested in the history of polar exploration.

I hope readers enjoy exploring these fifty objects. All bar four are presented broadly chronologically, in nine parts, which culminate in The Age of Aviation. The objects in the final part are presented in a slightly different format, allowing for wider consideration of issues relating to the conservation of historical objects.

For readers for whom *A History of Polar Exploration in 50 Objects* is a voyage of discovery, I wish them '*bon voyage*'. For those for whom some of the territory is already familiar, I hope you find some uncharted areas or an object which sheds new light on your own experience of polar regions or the history of polar exploration.

Anne Strathie
Cheltenham, 2024

Part I

Laying Foundations

James Cook was recognised during his lifetime as one of the British Royal Navy's greatest mariners and surveyors and is still regarded as such by many. During his first two circumnavigations he attempted to reach or, failing that, delineate what was known as the Southern Continent or *Terra Australis Incognita* ('unknown southern land'). On his third circumnavigation, his orders were, if possible, to navigate the fabled Northwest Passage from the Pacific Ocean eastwards.

Cook, who was raised in rural north Yorkshire, joined the navy in his late twenties, following an apprenticeship and a period working with Whitby-based shipowner John Walker. After joining the navy in 1755, Cook served in the Seven Years' War and, thanks to innate ability and encouragement and support from his commanding officer, Hugh Palliser, he was appointed Surveyor of Newfoundland. While Cook was in the latter post, his observations of a solar eclipse also brought him to the notice of members of the Royal Society.[1]

By early 1768, officials of the Royal Society were planning an expedition to the Pacific Ocean, during which Royal Society scientists would take readings of June 1769's transit of Venus over the sun. The Royal Society hoped Admiralty officials would loan them a vessel – which they did, on condition that a competent naval officer was in command. Although Cook was still only a non-commissioned officer, the highly competent 6ft-tall ship's master was promoted to lieutenant and given command of HMS *Endeavour*, a converted Whitby cargo ship.[2]

When *Endeavour* left Britain in the summer of 1768, Cook was accompanied by Royal Society nominees including astronomer Charles Green, naturalist Joseph Banks and the latter's entourage of scientists and artists. On 3 June 1769, on the Pacific Ocean island of Otaheite (present-day Tahiti), Green and Cook recorded the passage of Venus. With the first stage of his mission complete, Cook opened a second set of Admiralty orders, which instructed him to sail to 40°S where, according to the Royal Society's chief geographer Alexander Dalrymple, he might meet the northern shores of the great Southern Continent.

In summer 1771, Cook returned to London. While he had seen no signs of a southern continent, he had, by circumnavigating New Zealand, established that the latter was not part of the supposed continent. Cook had also charted the east coast of New Holland and, thanks to his experiments with a range of recommended anti-scorbutics, kept scurvy, the scourge of long-distance mariners, at bay during *Endeavour*'s three-year absence.[3]

After reporting to the Admiralty, Cook was promoted to commander and presented to the king. While still working on his expedition reports, he accepted a commission for a second circumnavigation.

Portrait of James Cook (lithograph based on Nathaniel Dance's portrait, *c.*1775); image courtesy of the Wellcome Collection.

1

HMS *Resolution* in Pack Ice

The watercolour 'Resolution in a Stream of Pack-Ice, 1772–3' was one of many works produced by artist William Hodges (1744–97) during and following James Cook's second circumnavigation. Cook believed that drawings, watercolours and paintings produced by Hodges would give those who saw them a more vivid impression than Cook's own written descriptions. During the voyage, Hodges produced many small watercolours, showing the ships among pack ice or icebergs, in the hopes of conveying something he had never seen before to those who were unlikely to ever see it themselves.

This wash and watercolour image, painted on laid paper (around 8½in square), is in the collection of the Captain Cook Museum, Whitby, England.

During his second circumnavigation, James Cook was given command of two converted Whitby-built vessels, *Resolution* and *Adventure*, and provided with a brand-new chronometer based on John Harrison's famous 'sea watch', which would enable him to calculate longitudes to much higher degrees of accuracy than previously.[1] Both Cook and the Royal Society hoped Joseph Banks would join the voyage, but the much-enlarged scientists' quarters that had been built to accommodate Banks and his team made *Resolution*'s deck unstable. After Admiralty officials insisted they were demolished, Banks resigned from the voyage and, with his Royal Academician artist friend Johann Zoffany, several scientists and John Gore (one of Cook's *Endeavour* officers), left London for Iceland.

When *Resolution* and *Adventure* left Britain in July 1772, Royal Society-appointed scientists Johann and Georg Forster and Admiralty-appointed artist William Hodges were aboard. At the Cape of Good Hope, the

'Resolution in a Stream of Pack-Ice, 1772–3', William Hodges;
image © and courtesy of Captain Cook Memorial Museum, Whitby.

Forsters 'botanised', while Hodges sketched and painted landscapes around Table Bay. As they continued south, the Forsters recorded sightings of penguins and unfamiliar seabirds, and Hodges made sketches of the ships passing through ice-scattered seas and among huge 'ice islands'.

Despite Cook spending hours on the main mast with his telescope, he saw no land to the south. He was, however, encouraged that ice chunks harvested from the sea, when melted down, produced fresh-tasting drinking water rather than thirst-inducing brine. As he continued, Cook

'Ice Islands', William Hodges (ref. kXReyb473GbXp); image courtesy of Mitchell Library, State Library of New South Wales, Sydney.

became increasingly convinced that, should the Southern Continent exist, it must lie south of 60°S.

On 17 January 1773, at around 39°E, *Resolution* and *Adventure* crossed the invisible Antarctic Circle. After sailing in rapid succession through clear water, loose ice and ice islands of assorted sizes, they passed 67°S and came up against what Cook recalled Greenlanders describing as 'field ice'. As close-packed ice fragments and ice islands up to 20ft high clustered round his ships, Cook climbed to the masthead in hopes of finding a way through.

To Cook's eye, some ice islands looked beautiful, while others appeared threatening – qualities he hoped Hodges might convey more eloquently with pencils, brushes and paints than he, a self-confessed 'plain-speaking' Yorkshireman, could in words. As the southern winter approached, Cook and his shipmates admired the ephemeral beauty of celestial displays which, as they reminded Cook of northern Aurora Borealis, he called Aurora Australis.[2]

Cook returned to New Zealand where, in safe harbour, his ships could be overhauled – and Hodges could paint *Resolution* at anchor and work up his sketches of ice formations. In June, Cook decided to escape the southern winter and return to Otaheite to replenish the water on his ships and restock with fresh produce. On the way back to New Zealand, the smaller, slower *Adventure* became separated from *Resolution*, leaving Cook with little alternative but to suggest that her captain, Tobias Furneaux, sail her back to England, leaving him and those on *Resolution* to search for the Southern Continent.

The so-called southern summer barely merited the name as, during December, *Resolution* was engulfed by blizzards, while icicles formed on the rigging and men's noses. Shortly after *Resolution* passed 67°S she became ice-beset, so Cook followed the pack edge to 70°S, when he saw open water 'leads' among the pack ice and turned south again.

On 3 February 1774, at 107°W, *Resolution* reached 71° 10'S. As Cook scanned the horizon, he longed to continue south. He concluded, however, that with 100 ice islands in view and daylight hours reducing fast, it would be a 'dangerous and rash enterprise' to continue.[3] He remained unsure whether the Southern Continent existed, so suggested in his reports that either the pack ice extended to the South Pole or the Southern Continent, from which ice islands probably emanated, was smaller and lay further to the south than was currently believed.

While Cook and his men saw many natural wonders during the southern leg of their circumnavigation, they were glad to return to the Pacific Ocean, where fresh produce was plentiful, and they made a first visit to Rapa Nui, or Easter Island, where they saw huge statues of the islanders'

'The Ice Islands' [9 January 1773], William Hodges, showing men collecting ice for fresh water (ref. 4B3E2wzJO2vjy); image courtesy of Mitchell Library, State Library of New South Wales, Sydney.

ancestral chiefs. In mid-January, after rounding Cape Horn and entering the south Atlantic Ocean, Cook saw, at around 54°S 38°W, a range of mountains, which appeared to be covered by snow and ice.[4] On 17 January he landed and, after firing a few rifle volleys, claimed the territory in the name of King George; the only living creatures much in evidence were fur seals and the largest penguins Cook or the Forsters had seen. When they reached and rounded a headland, however, Cook realised he was surveying an island rather than (as he had hoped) the northernmost reaches of the elusive Southern Continent. After charting the offending headland as 'Cape Disappointment', Cook named the island for King George and headed north.[5]

In late summer 1775, Cook returned to London, where he tried to summarise his findings for his reports. If there was a Southern Continent, he explained, it was 'doomed by nature never once to feel the warmth of the sun's rays, but to lie buried in everlasting snow and ice'. Should other mariners venture there, they would probably find 'thick fogs, snowstorms, intense cold, and every other thing that can render navigation dangerous'. And should they venture further south than he had, they risked 'being fixed there forever, or of coming out in an ice-island'.

The Admiralty's contract with Hodges suggested that his illustrations should 'give a more perfect idea than [could] be formed from written descriptions only'. Admiralty officials were so pleased with Hodges's portfolio of works that they asked him to prepare engravings of a selection of them for inclusion in the printed expedition reports, so readers would have an impression of the wonders Hodges, Cook and their shipmates had seen.

POLAR POSTSCRIPT: Cook's circumnavigations established a template for expeditions, which combined geographical exploration and surveying with detailed scientific enquiry and, at the conclusion of the voyage, published reports illustrated with maps, charts and engravings made from drawings and paintings by professional or highly skilled amateur artists.

2

Elizabeth Cook's Ditty Box

This 'ditty box'* was presented to Mrs Elizabeth Cook after her husband's ship, *Resolution*, returned to London in 1780. The oak box was made by crew members from timbers from the ship. It was carved with Polynesian symbols and decorated with miniature silver plates etched with inscriptions relating to the box and Cook's career.

The ditty box, measuring around 3½in by 2¾in, is in the Dixson Library collection in the State Library of New South Wales, Sydney, Australia (ref. SAFE/DR2; npAd3Ob1).

In summer 1775, Elizabeth Cook was, as ever, relieved to welcome her husband James back to their home following another long voyage.[1] In his absence, she had mourned the death of baby George, the third of their five children to die while he was away. She was, however, gratified when he was promoted to the rank of captain and offered the sinecure post of Fourth Captain of Greenwich Hospital, which provided him with an income while he wrote up his expedition reports. She was also pleased when he was elected to Fellowship of the Royal Society and invited to submit a paper on his use of anti-scorbutics as a means of keeping scurvy at bay during his circumnavigations.

Although Cook enjoyed attending Royal Society meetings with his wife, he soon began spending increasing amounts of time with his friend and mentor Hugh Palliser, who was now a member of the Admiralty Board. Palliser told Cook that Joseph Banks's friend Constantine Phipps

---

* Ditty boxes were made and used by mariners to store small, important personal possessions during voyages.

Mrs Cook's ditty box showing decorative features and multi-part cover (ref. npAd3Ob1); images © and courtesy of Dixson Library, State Library of New South Wales, Sydney.

(who had family estates near Whitby) had recently failed to reach the Pacific Ocean via Spitsbergen, the putative Open Polar Sea and the Bering Strait.[2] As the Admiralty Board were now planning a Northwest Passage expedition, starting from the Pacific Ocean and Bering Strait, Palliser wondered if Cook could help him and his fellow board members with drawing up a shortlist of potential commanders for such an expedition.

In January 1776, following an informal dinner with Palliser and other Admiralty Board members, Cook agreed to lead the forthcoming expedition. While Elizabeth prepared for the arrival of another child, Cook was informed that he would again command *Resolution*, while Charles Clerke (who had served on his last two circumnavigations) would captain *Discovery*, another Whitby-built vessel.

In late May, Elizabeth – shortly after attending a Royal Society meeting with her husband – gave birth to a son, whom they named Hugh for his godfather, Hugh Palliser. On 12 July, Cook sailed from Plymouth.

Shortly after Elizabeth learned of her husband's safe arrival at Cape of Good Hope, she was presented, on his behalf, with the Royal Society's Copley Medal for his paper on scurvy.[3] Although the Cooks' elder sons, James and Nathaniel, were also now at sea, Elizabeth had Hugh for company and passed the time by making an embroidered waistcoat for her husband from *tapa* fabric he had brought back from Otaheite.[4] She had little idea when to expect her husband back, so was shocked when, in early January 1780, newspapers reported that he had died in February 1799 on Owhyhee, a remote Pacific Ocean archipelago.[5]

In due course, Elizabeth learned that her husband had first visited Owhyhee in January 1778 to rewater his ships before exploring and charting America's north-west coast. After finding no likely route running from the west coast to Hudson's or Baffin Bays, he had continued to the Bering Strait and across the Arctic Circle. When his ships became ice-beset at around 70°N, he had returned south and, rather than overwinter in chilly Kamchatka, returned to Owhyhee. Although the islanders had initially seemed as friendly as before, a series of misunderstandings resulted in a skirmish during which Cook, four marines and sixteen islanders died.

Charles Clerke and Lieutenant John Gore, now in charge of the expedition, decided to return to the Bering Strait the following year. They were soon blocked by ice, and as Clerke was now seriously ill, headed for Kamchatka. Clerke died before they reached there so, after dispatching reports to London, Gore captained Cook's ships back to Britain.[6]

King George, who reportedly wept on learning of Cook's death, granted Elizabeth a pension and financial support for her three sons. When Cook's ships returned, Elizabeth was presented with a carved oak ditty box, fashioned in the shape of a coffin and decorated with patterns similar to those on the *tapa* fabric she had been using to make the waistcoat her husband would never wear. Set in the wood were four silver discs made from beaten coins, each bearing an inscription: 'Made of Resolution oak for Mrs Cook'; 'Captain James Cook slain at Owhyhee, 14 February 1779'; 'Quebec Newfoundland Greenwich Australis'; and 'Lono and the Seaman's idol' – the last being, Elizabeth learned, a reference to a religious ceremony celebrated in Owhyhee.[7] Within the ditty box lay a lock of Cook's hair, a watercolour of Owhyhee and other mementos.[8]

Within months of her husband's ships returning, Elizabeth learned that her 15-year-old son Nathaniel had drowned when his ship, HMS *Thunderer*, went down with all hands during a Caribbean hurricane.[9] While she grieved, her husband's posthumous fame grew, thanks to eulogistic poems and prose, printed reproductions of portraits by Royal Academician Nathaniel Dance and others, and a Royal Society commemorative medal struck at the request of Joseph Banks, the society's then president. Banks also encouraged Elizabeth to collaborate with Royal Society member Andrew Kipps on a biography of her husband, which was published in 1788, with Dance's portrait as a frontispiece. That year, Elizabeth and her younger son Hugh moved to a spacious new home in Clapham's High Street, conveniently close to Merton, where her cousin Isaac Smith (who had regularly sailed with Cook) had a country home to which he often invited Elizabeth and Hugh.

In 1793, 16-year-old Hugh left Clapham to study theology at Christ's College Cambridge and his brother James, now approaching 30, received his first naval command. Within a year, however, Elizabeth lost her two remaining children after Hugh succumbed to a fever in his college rooms and James died in an accident when returning to his ship.[10] Although Elizabeth was near-paralysed by grief, she gradually, thanks to Isaac Smith and other relatives, came out of mourning and resumed her active social life.

In 1835, Elizabeth died in her Clapham home, aged 94. The woman described in newspapers as the 'esteemed and respected [...] widow of the celebrated circumnavigator' retained her mental faculties to the end. Her husband's expedition records and charts were already in safekeeping with the Admiralty and Royal Society and, shortly before she died, she sent her copy of the Royal Society's commemorative medal to the British Museum for safekeeping.[11] As she had regarded her husband's public roles as separate from their private life, she destroyed his personal letters to her and gave his gifts and other personal mementos, including her precious ditty box, to relatives and close friends who had supported her over her long life.

POLAR POSTSCRIPT: In 1806, Elizabeth Cook gave her ditty box to a relative, John Carpenter, from whom it passed by descent to Thomas Hart. It was later purchased by Sir William Dixson, who bequeathed it, along with other items from his collection, to the State Library of New South Wales. Although there are numerous memorials to Elizabeth's husband in public places in Britain and elsewhere, her will made provision for a memorial to the entire Cook family to be made and erected in the Church of St Andrew the Great, in Cambridge, where she and her sons Hugh and James are buried.

3

Scoresby's Barrel Crow's Nest

During summer 1807, whaling master William Scoresby of Whitby used, for the first time, a mast-mounted 'barrel crow's nest' of his own design. He could now, with both hands free, use his telescope to search the horizon for whales' spouts and leads through the Arctic sea ice. The barrel crow's nest, sometimes referred to as Scoresby's 'tunna' (from '*tun*', meaning barrel), soon became a standard feature on British whalers and other vessels, particularly those operating in polar regions.

This replica crow's nest (approximately full size) is based on Scoresby's design and is in the Whitby Museum. The museum also holds documents and other artefacts relating to William Scoresby and his son William, who became both a whaler and respected scientist.

William Scoresby, like James Cook, was raised in rural Yorkshire and moved as a young man to Whitby, where he served a mariner's apprenticeship with relatives of Cook's employer and mentor, John Walker.[1] In 1790, Scoresby's first season in command of the whaling ship *Henrietta*, he and his employer were both disappointed with his catch. The following year, however, when there was more sea ice, he returned with a record catch of eighteen whales which, thanks to high demand for whale oil and ribs, meant he was well paid for his efforts.[2]

By 1806 Scoresby owned his own whaler, which he named for James Cook's *Resolution*. During her first season, Scoresby and his 16-year-old son William reached 81° 30'N, a new 'Farthest North' for British vessels operating off Greenland. That year, the Scoresbys returned with over twenty whales, as well as seals and walruses and – to the amazement of Whitby's residents – two white-coated Arctic bears.

Replica of a crow's nest designed by William Scoresby Snr; collection of Whitby Museum; photograph © A. Strathie, with permission of Whitby Museum.

Over the winter, Scoresby designed an apparatus which would allow ships' captains or other lookouts to remain safe when aloft, rather than holding masts and telescopes with frost-nipped fingers.[3] Scoresby's barrel crow's nest, mounted on the main mast, could be entered through a trapdoor in its base. The interior was fitted with internal racks and hooks where telescopes, instruments, signal flags and a loud hailer could be stored when not in use. In recognition of the vagaries of Arctic weather, Scoresby also fitted the crow's nest with a retractable hood, which protected the user and telescope lenses from rain, sleet and snow. During the 1807 whaling season, Scoresby and his son William – who had joined his father as a 10-year-old but was now *Resolution*'s chief officer – made good use of the new crow's nest.

Although William's formal schooling had taken second place to his father's whaling activities, he began spending winters in Edinburgh, studying natural history, mathematics and logic with Professor Robert Jameson and other university tutors. William's education was again interrupted when he and several friends were enticed to enrol for the navy, which was looking for new recruits to fight the French. But his new career ended abruptly after a near shipwreck on his first voyage resulted in him being discharged at Portsmouth.[4] When William told his father where he was, his father asked him to make a detour via London, where he had left reports he wanted delivered to the Royal Society's President, Sir Joseph Banks.[5] Banks was interested that William was studying in Edinburgh with Robert Jameson (whom Banks knew) and invited William to keep in contact.

Following another summer in *Resolution*'s crow's nest, William returned to Edinburgh, where he completed his studies and was elected to membership of the city's Wernerian Society. In 1811, William married and, after his father moved to Scotland, he became a Whitby whaling master in his own right. Despite his additional responsibilities, he continued his scientific work, of which he sent details regularly to Jameson.

William also kept up his correspondence with Joseph Banks, who kindly sent from London a new, highly accurate, deep-sea marine thermometer. When William noticed in summer 1817 that there was less ice between 74°N and 80°N than in previous years, he wrote to advise Banks, who suggested this tallied with reports of melting alpine glaciers and other indications that, following centuries of historic lows, northern hemisphere air temperatures appeared to be rising.[6] After William sent Banks a copy of his 'Treatise on the Northern Ice', Banks reminded him that a 1745 Act of Parliament offered 'a reward of £20,000 for the discovery of a NW passage and £10,000 for the ship that shall first reach the 89th degree of North Latitude'.

When William explained to Banks that whalers were neither equipped for nor could afford to embark on such speculative ventures, Banks invited him to London to meet Admiralty officials who were organising an Arctic 'voyage of discovery'. Banks had sent a copy of William's treatise to Admiralty officials, although, when William met John Barrow, the Admiralty's apparently influential Second Secretary, the latter made no mention of the paper and told William that only naval officers were eligible for senior expedition posts.[7] After Banks warned William that Barrow was unlikely to change his mind, William spent the winter working on a book Jameson wanted him to write on the Arctic regions.[8]

In early 1818, Banks confirmed that the Admiralty had appointed Captain David Buchan and Lieutenant John Franklin to lead an

HMS *Alert* crow's nest, from *Illustrated London News*, 5 July 1875; author's collection.

expedition to Spitsbergen and the North Pole on *Dorothea* and *Trent*, while Commander John Ross and Lieutenant Edward Parry, on *Isabella* and *Alexander*, would survey Baffin Bay, then attempt to enter the Northwest Passage.[9] Parry was also charged with locating the North Magnetic Pole which, as Admiralty officials and mariners knew, increasingly affected the accuracy of ships' and other compasses as they approached it.

In mid-April 1818, while heading north to the whaling grounds, William wrote to Banks from Shetland. Among other matters, he reminded Banks that, notwithstanding the public's and the Admiralty's expectations, there was no guarantee that the approaching summer would find northern waters as relatively ice-free as they had been in 1817. Should ice conditions be more like 1816 and previous years, William would, he admitted, be surprised if Buchan's party passed 84°N or Ross's men traversed the Northwest Passage on their first attempt.

In the event, William's prediction proved correct. He was running ahead of Buchan and Franklin's ships and, despite his best efforts from the crow's nest, became ice-beset at around 80°N. With more ice than the previous year, William returned to Whitby with a good catch. A few months later, he learned that *Dorothea* and *Trent* had also been forestalled at around 80°N and that Ross's ships had failed to enter the Northwest Passage.

POLAR POSTSCRIPT: Crow's nests based on William Scoresby's 1807 design were widely used throughout the nineteenth and early twentieth centuries by whalers and sealers and on Admiralty and other ships exploring polar regions.

4

A Panorama of Spitsbergen

During Commander David Buchan's 1818 expedition to Spitsbergen and the North Pole, Lieutenant Frederick Beechey (son of Royal Academician Sir William Beechey) produced panorama drawings of the area north of Spitsbergen where expedition ships *Dorothea* and *Trent* became frozen in.[1] Beechey's drawings were later scaled up and used to create a painted panorama which was shown at Henry Barker's Rotunda off Leicester Square, London. Barker's father was Scottish panorama artist Robert Barker, who had pioneered panoramas as a form of popular entertainment, particularly for audiences interested in seeing large-scale images of distant lands.[2]

No original canvases of the Spitsbergen panorama exist. Keyed drawings of the panoramas, as shown here, were incorporated into guides which were sold to visitors. Examples of them can be found in archives and libraries and are occasionally offered for sale.

On 11 April, Easter Monday 1819, customers arrived at Henry Aston Barker's Rotunda off Leicester Square, eager to see what was advertised as a 'novel scene' from a recent polar expedition. After paying a shilling to enter, customers mounted a circular platform from where they viewed a long, 30ft-high mounted canvas panorama of an Arctic vista. A keyed diagram and text explained that HMS *Dorothea* and HMS *Trent* (both visibly listing) were trapped in the ice and their captains, David Buchan and John Franklin (seen in the foreground), were 'consulting together on future proceedings'.

Visitors' attention was also drawn to the magnificent snow-capped mountains and huge ice barrier, and to polar bears, walruses and other

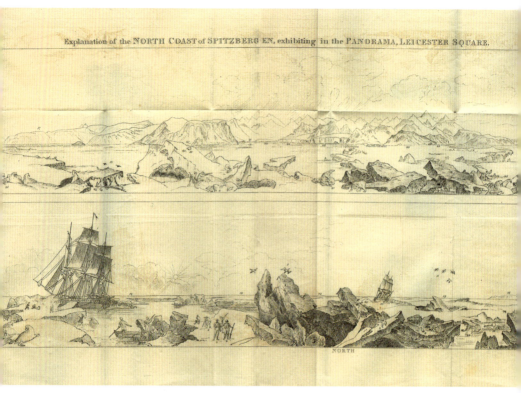

'Panorama of the North Coast of Spitzbergen', Henry Aston Barker (from Beechey's drawings); image © and courtesy of Russell A. Potter.

unfamiliar creatures. While the awe-inspiring scenery recalled scenes in *Frankenstein*, a recent novel, or Coleridge's *Rime of the Ancient Mariner*, the magnificent and apparently peaceful vista made it difficult for visitors to imagine Buchan, Franklin and their men winching boats to just beyond 80°N or being tossed in storm-damaged ships in ice-strewn seas.

Although Barker's Rotunda could accommodate two panoramas, there was no companion display showing Baffin Bay and Lancaster Sound, which John Ross and Edward Parry had explored on HMS *Isabella* and *Alexander* during a second Admiralty expedition. John Barrow had no desire to publicise this second expedition, notwithstanding that John Ross had, in accordance with orders, completed the first survey of Baffin Bay since William Baffin first charted it 200 years previously. Ross had also, with assistance from Royal Society representative Edward

A Panorama of Spitsbergen

Barker's Rotunda, Leicester Square, cross-section from Robert Mitchell's *Plans and Views in Perspective …* (pub. 1801) (ref. 52.519.153); Rogers Fund 1952, Metropolitan Museum of Art, New York (Open Access Initiative).

Sabine and others, taken and recorded numerous magnetic and other readings. He had established that Cumberland Sound, one of several channels leading off Baffin Bay, was an impasse, and, as instructed, entered Lancaster Sound. After Ross entered the sound and saw what he thought were mountains blocking his way west, he had, to Edward Parry's disappointment, turned back well before reaching the uncharted mountains – which Ross designated 'Croker's Mountains' in honour of the Admiralty's First Secretary.

Barrow wrote a scathing review of Ross's official expedition report; when the review was published anonymously it resulted in Ross being satirised and caricatured.[3] And although Ross had lost no ships and could not be court-martialled, Barrow ensured Ross was summoned to an Admiralty hearing. So vociferous was Barrow's criticism of Ross's actions that Ross found it difficult to persuade his nephew James (who had sailed with him since boyhood) to testify in his favour. After Parry and the Admiralty's magnetism expert Edward Sabine spoke out against Ross, John Barrow effectively debarred Ross from further expeditions.

Shortly after the Spitsbergen panorama opened, Edward Parry, now in command of HMS *Hecla* and *Griper*, left Britain with Beechey, Sabine, James Ross and other veterans of the 1818 expeditions to embark on a second attempt at traversing the Northwest Passage. While they were away, John Franklin would lead a small party, including *Trent* veteran George Back and naval surgeon-cum-naturalist Dr John Richardson, to Hudson's Bay. From there, they would be joined by experienced Hudson's Bay Company (HBC) personnel, who would accompany them down the Coppermine River to the north Canadian coast where, all being well, they would join Parry's ships and complete the Northwest Passage.

Meanwhile, John Ross remained in Britain on half pay and mourned the death of his baby daughter, who had died in his absence. In September, he and his wife welcomed a baby son, Andrew. By then, his expedition report – lavishly illustrated with drawings and paintings by him and his expedition's Inuit translator John Sacheuse – had run to a second edition. As the panorama based on Beechey's drawings continued to draw crowds, Ross provided panorama proprietors Messrs Marshall with some of his illustrations, which their painters used as templates for large canvas panels within one of Marshall's famous 'peristrephic', or moving, panoramas, which regularly toured Britain.

By April 1820, Marshall's artists had produced, from Ross's work and a further selection of Beechey's drawings, a 'Grand Panorama of the Magnificent Scenery of the Frozen Regions'.[4] The panorama opened at Birmingham's Shakespeare Tavern, then toured the country. Performances were regularly enhanced by live music, 'brilliantly illuminated' evening performances and an Arctic 'museum', which featured a stuffed white bear and other Arctic animals, a 15ft kayak and examples of Inuit clothing that Ross had brought back from his voyage. By October, when Parry and his men returned to London, the original Spitsbergen panorama had completed its run, but Marshall's 'Frozen Regions' panorama was still attracting crowds in Edinburgh.

Parry had, Ross learned, reached Lancaster Sound early in the season, sailed straight through the area of the map where Ross had marked Croker's Mountains and continued to 110°W before becoming blocked by huge quantities of ice flowing south into what Parry had named Barrow

Strait. For Barrow, Parry was now the man of the moment and Barrow already had plans for him – in the form of an expedition to Hudson's Bay, from where Parry was instructed to find a southern route which would bypass the heavy ice which had prevented *Hecla* and *Griper* from reaching the Bering Strait and Pacific Ocean.

In January 1823, while Marshall's 'Frozen Regions' panorama continued to tour Britain's major towns and cities, the *Bath Chronicle* announced its imminent arrival in the home town of 'Capt. Parry, the most heroic and intrepid voyager that ever crossed the ocean'.[5] The 'intrepid voyager' was still in the Arctic, but those looking forward to the opening of 'Frozen Regions' also had to contend with adverse weather conditions. As the *Chronicle* informed its readers, the freezing of canals between Birmingham and Bath would result in the late arrival in Bath of the 'Frozen Regions' panorama. The panorama was, potential visitors were assured, well worth waiting for, as the Bath performances featured live musical accompaniment, including perennial favourites such as 'Hearts of Oak' and the specially commissioned 'Captain Parry's Waltz'.

POLAR POSTSCRIPT: Panoramas, including of polar and other distant regions, remained popular with the public until the 1890s. Barker's Rotunda was purchased in 1865 by the Archdiocese of Westminster and converted into a church, of which the Rotunda's dome remains a notable feature.

5

William Scoresby's Manuscript

William Scoresby, at the suggestion of Professor Robert Jameson of Edinburgh University, wrote his first book, *An Account of the Arctic Regions, with a History and Description of the Northern Whale-Fishery*, between whaling seasons. It was published in 1820 in two volumes (around 600 pages each) by William Constable of Edinburgh and praised as the first comprehensive survey of the Arctic region. The publication of his book resulted in Scoresby being recognised as a polar scientist as well as a successful whaling captain.

The manuscript and Scoresby's first-edition copy of his book are in the Whitby Museum, Whitby, Yorkshire.

When William Scoresby's whaler became ice-beset during the summer 1818 season, he took magnetic and other readings, made surveys and wrote descriptions of Spitsbergen's glaciers and sea ice, all of which he planned to incorporate into his forthcoming book on the Arctic regions. Scoresby's mentor, Professor Robert Jameson, introduced him to his publisher, Archibald Constable, who, while best known for publishing Walter Scott's novels, also published scientific textbooks by Jameson and others.[1] In December, following the birth of Scoresby's second son, Constable confirmed that his company would publish his manuscript in two illustrated volumes, and that, six months after publication date (probably spring 1820), they would pay him £250 for his efforts.

In May 1819, Scoresby, accompanied by his wife and two sons, moved to Liverpool, where he juggled working on his manuscript with overseeing the construction of a new whaler which his new business partners had promised would be ready for the 1820 season. As the result of Edward

William Scoresby's Manuscript

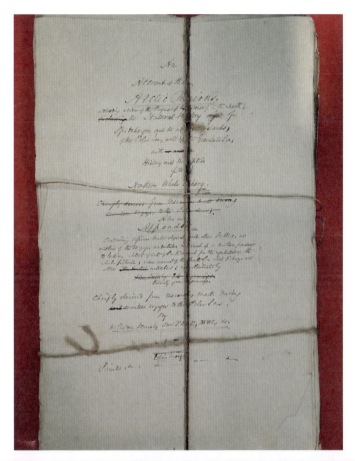

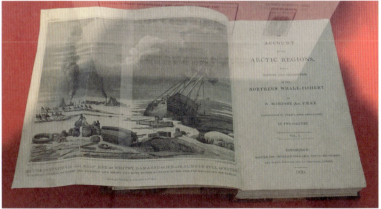

Manuscript of Rev. William Scoresby's *An Account of the Arctic Regions*.
First edition of *An Account of the Arctic Regions*, published by William Constable (1820).
Both photographs © A. Strathie, taken in display cases with permission of Whitby
Museum; the manuscript is copyright Whitby Museum.

Parry's second expedition was unlikely to be known before his book went to press, Scoresby's chapter on the Northwest Passage was largely historical in nature. As a proud Yorkshireman, he praised Whitby's 'adventurous navigator', James Cook, for his attempt to traverse the passage from the Bering Strait. He also suggested that, despite John Ross's failure to enter the Northwest Passage from Lancaster Sound, his 'brief and unostentatious' expedition report and his and Parry's charts of Baffin Bay were comprehensive.

As Scoresby's book was likely to outlast the Croker's Mountains controversy, he showed Lancaster Sound as an open-ended channel from which a speculative 'Frozen Ocean' continued to the Bering Strait. As to the feasibility of traversing the passage, Scoresby noted that he considered it unlikely that the passage could be completed in a single summer and ships of 100–200 tons were more likely to pass through the passage's icy, sometimes narrow channels than larger vessels – although he conceded the latter could carry more provisions.[2] He also praised the dog sledge-driving capabilities of HBC employees, Inuit and other inhabitants of Arctic regions.

Scoresby was not, he admitted, a believer in the concept of an 'Open Polar Sea' beyond the northern Arctic pack ice, but he accepted that, given Parliamentary rewards were available for reaching the North Pole or Bering Strait by that route, the search for it would continue.

Scoresby supplemented his historical overviews with chapters on a myriad of topics, from seals to snowflakes and scurvy to sledging. He also included detailed meteorological and other tables and charts and illustrations. In a dedication, he acknowledged the 'early and uniform friendship' of Professor Jameson – whose nomination of Scoresby to Fellowship of the Royal Society of Edinburgh allowed Scoresby to append 'F.R.S.E.' to his name on the title page. Scoresby also thanked Joseph Banks for 'friendly suggestions and encouragement' and his 'kindness and liberality' in loaning and presenting him with scientific instruments and other items.

In mid-March 1820, following the publication of *An Account of the Arctic Regions*, Scoresby received a package of author's copies from Constable, which he distributed to Jameson, Banks, his father and others. That done, he sailed for Greenland on his new ship, *Baffin*. On the way north, he stopped at Stranraer in hopes of recruiting additional crew members and was pleased to meet an acquaintance, David Gordon, who invited Scoresby to spend the evening with him and his friend, John Ross.[3]

Fig. 2 from Scoresby's paper to the Royal Society of Edinburgh (18 December 1820) showing 'remarkable Atmospheric Reflections and Refractions, observed in the Greenland Sea'; image A. Strathie.

Scoresby and Ross soon found common ground (other than being the object of John Barrow's apparent disdain) and Scoresby gladly accepted Ross's invitation to visit his lochside home the following day. Scoresby enjoyed seeing Ross's manuscript charts of Baffin Bay and, by way of thanking Ross for gifts of printed charts and a copy of his pamphlet on marine instruments, he invited Ross to lunch on *Baffin* the following day – when Ross presented Scoresby with a copy of his 1818 expedition report. That evening, following a farewell dinner at Ross's home, the two independently minded polymaths agreed to remain in contact.

Baffin performed well on her maiden voyage and, although Scoresby captured no whales until mid-May, he returned to Liverpool in late August with what he learned was the largest catch ever landed there from Greenland waters.

Back in Whitby, he found his wife and two young sons well, although he was saddened to learn of the death of his mentor and supporter Joseph Banks. On a more positive note, his publishers confirmed they were 'much pleased' with sales of *An Account of the Arctic Regions*, and with reviews that praised it for its blend of scientific and practical information.

Scoresby had included a short section on atmospheric refractions in his book, but his new-found friendship with Ross made him more interested in the subject. As he had recently seen refractions which, like those Ross had seen in Lancaster Sound, resembled a mountain range, he submitted a paper on the topic to the Royal Society of Edinburgh. When the paper was accepted, Scoresby felt he had both supported his friend John Ross and enhanced his own growing reputation as a fully fledged scientist.[4]

POLAR POSTSCRIPT: Scoresby's *An Account of the Arctic Regions* became an indispensable standard text for polar scientists and explorers and remained so for over a century. In more recent years, Scoresby's name became familiar to readers of Philip Pullman's *His Dark Materials*, through the character Lee Scoresby who, appropriately, spends time in the Arctic.

Part II

Exploring North and South

John Barrow knew that many of the men who volunteered for the current wave of Admiralty 'voyages of discovery' to the Arctic and other regions were interested in exploring uncharted areas or learning more about magnetism or other areas of science. There was no denying, however, that the prospect of returning to full pay and active service after the enforced idleness of the post-Napoleonic Wars era was also an attraction. There was also, as Edward Parry and naval officers knew, the possibility of advanced promotion or a share of Longitude Discovery Bill awards ranging from £5,000, for reaching 110°W or 89°N, to £20,000, for charting a route between the Atlantic and Pacific Oceans, whether via the Northwest Passage or the North Pole.

As William Scoresby had explained to Joseph Banks and John Barrow in 1817, Arctic whalers and other mercantile mariners could not afford to forgo a season's earnings on the off-chance of qualifying for a Parliamentary award. But as mercantile mariners, sealers and whalers passed through uncharted waters, some carried out their own investigations. One such was ship's master William Smith, who when transporting cargo from Buenos Aires to Valparaiso in 1819, was swept south to around 62°S and saw unfamiliar islands and an abundance of seals and whales. The following year, he and Edward Bransfield of the British Navy's station in Valparaiso returned to the islands, which they charted as New South Shetland.[1] While returning to Valparaiso, they saw more uncharted islands and, across the ice, what Bransfield described as 'high mountains covered with snow' rising from what he marked on his chart as 'Trinity Land'.[2]

Around the same time, Vice Admiral Fabian Thaddeus von Bellingshausen of the Russian Navy – a great admirer of James Cook – embarked on his second circumnavigation. This time, he was commissioned to investigate southern waters and, if possible, find and land on the great Southern Continent. On his way south from Russia, Bellingshausen visited and consulted with Joseph Banks and, when he embarked on his circumnavigation, concentrated his investigations in areas where Cook had not sailed below the Antarctic Circle.

In late January 1920, at around 69°S 2°W, Bellingshausen looked south and saw ice fields stretching to the horizon, and what he thought might be a range of snow-covered mountains.[3] Bellingshausen could not be sure whether they were mountains, or indeed if they were part of the Southern Continent, but he had, like Cook, narrowed down the area of enquiry and also, like his hero, been privileged to witness the Aurora Australis, one of the world's great natural phenomena.

6

Weddell's 'Sea Leopard of South Orkneys'

James Weddell, a sealer from Leith in Scotland, encountered the 'sea leopard' during his 1822–24 voyage to the sea which now bears his name. During the voyage, Weddell passed James Cook's Farthest South and identified what he thought might be a new species of mammal. The caption to the drawing suggests it was 'drawn from nature', but the unusual proportions suggest a stuffed specimen may also have served as a model for the final version.

The image was published in Weddell's *A Voyage Towards the South Pole, Performed in the Years 1822-24* (Longman et al., 1825–27), copies of which can be found in libraries and archives.

Scottish mariner James Weddell joined the Royal Navy as a boy, but following his training, he served on both naval and mercantile vessels. Like many other Royal Navy seamen, he was paid off following the Napoleonic Wars and rejoined the mercantile service. In late 1819, Weddell was in the Falkland Islands on a 150-ton brig, *Jane*, when he heard reports that William Smith and others had identified potential new sealing grounds south of Tierra del Fuego and Cape Horn. By the time Weddell reached the South Shetland Islands, other British and American sealers were already operating there, but there was no shortage of fur seals and sea elephants, and he returned north with full holds.

In September 1822, Weddell set out from Britain on *Jane*, this time accompanied by a 65-ton cutter, *Beaufoy*. When he reached the South Shetlands, there were so many sealers competing for an apparently depleted seal population that he headed for more remote islands which British sealer George Powell and Nathaniel Palmer of Connecticut had

Sea Leopard of South Orkneys, from James Weddell's *A Voyage Towards the South Pole ... 1822–24*; image © and courtesy of Linda Hall Library of Science, Engineering & Technology.

visited the previous year. As Weddell charted the more remote islands, which lay round 60°S, he named them the South Orkneys for the Scottish archipelago at 59°N. As it seemed likely that the fur seal populations of both South Shetlands and South Orkneys might soon be permanently depleted, Weddell continued south and, thanks to favourable weather and calm, relatively ice-free seas, soon crossed the Antarctic Circle.

On 17 February 1823, Weddell passed 71° 10'S, the long-standing Farthest South set by James Cook. Three days later, at around 34°W, *Jane* and *Beaufoy* reached 74° 15'S. The ships were still in open water and, as there were few icebergs and no pack ice in sight, Weddell wondered whether to continue south. As there was also no land in sight, however, Weddell decided that as his rations were running low and daylight hours were reducing fast, he should return north.[1]

Weddell and his men were disappointed not to have seen and perhaps brought back cargo from the Southern Continent, but before returning north, celebratory grog was dispensed, flags were raised, a gun was fired, and Weddell announced to his men that they were now in 'The Sea of George the Fourth'. After a brief stop at South Georgia, Weddell overwintered in the Falklands before returning to the South Shetlands and South Orkneys for another sealing season. While there, he noticed

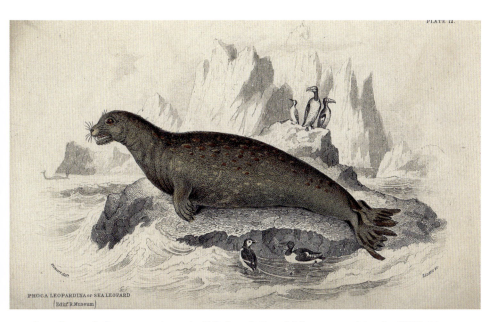

A sea leopard, from a specimen in Edinburgh Royal Museum, etching by W.H. Lizars after J. Stewart (pub. *c.*1839); image courtesy of the Wellcome Collection (no. 40712i).

an unfamiliar type of seal, a sea leopard with mottled grey and white fur, which he noticed seemed to prefer ice that was attached fast to the coasts of islands. Curious as to whether he had identified a new species, he retained some specimens so he could sketch them and would have something to show natural scientists when he returned to Britain.

By late summer 1824, Weddell was back in London, where his business partners, James Strachan of Edinburgh and John Mitchell of London, agreed to assist with the costs of publishing accounts of his recent seasons' voyages. When the first edition of *A Voyage Towards the South Pole, Performed in the Years 1822–24* appeared in 1825, it was illustrated with high-quality plates, including some of Weddell's drawing of the 'Sea Leopard of the South Orkneys' and an illustration showing *Jane* and *Beaufoy* in open water at their record southerly latitude.

Back in Scotland, Weddell offered a specimen of his 'sea leopard' to Edinburgh University, where it was examined by William Scoresby's mentor, Professor Robert Jameson, and others. The formal process took some time, but in 1826, Weddell's 'sea leopard' was classified as *Leptonychotes weddellii*, which soon became known as the Weddell seal.[2] In 1827, in recognition of his scientific work, Weddell was elected a Fellow of the Royal Society of Edinburgh.

Jane and *Beaufoy* in the ice, from Weddell's *A Voyage Towards the South Pole ... 1822–24*; image © and courtesy of Linda Hall Library of Science, Engineering & Technology.

POLAR POSTSCRIPT: Weddell was aware that, had he continued south, he might have sighted or landed on the coast of the Southern Continent. After his proposal to the Admiralty for an Antarctic expedition was rejected, he continued his mercantile career. Weddell died in 1834, the year after the Royal Geographical Society awarded their second Founder's Medal to sealer John Biscoe, whose employers, Enderby Brothers, funded a circumnavigation during which Biscoe charted Enderby Land, Graham Land and other areas of Antarctica.[3] In 1900, the ocean where Weddell identified the circumpolar seal that was named for him in 1826 was renamed the Weddell Sea in his honour.

7

Edward Parry's Deck Watch

Parry's handsome silver deck watch and its mahogany case were made in 1823–24 by Parkinson & Frodsham of London. The deck watch was procured by the Admiralty (as evidenced by the 'broad arrow' symbol) and issued to Edward Parry for use on his fifth Arctic expedition. Deck watches, chronometers and other timepieces used to measure longitude were made to high standards of accuracy and were regularly checked during voyages, including against each other, to ensure they had not stopped or begun to gain or lose time.

The deck watch is in the collection of the Worshipful Company of Clockmakers (ref. 447) at the Science Museum, London.

Edward Parry had joined the Royal Navy at the age of 13, so he well understood the importance of timepieces in measuring longitude. When he was in command of expeditions to the Northwest Passage, however, the accurate measurement of longitude became particularly significant as it would determine his and his shipmates' eligibility for Parliamentary awards of up to £20,000. In May 1819, when Parry and his men left London for the Northwest Passage on *Hecla* and *Griper,* they carried with them twelve chronometers and timepieces made by Parkinson & Frodsham and other long-standing Admiralty suppliers.[1]

On entering Lancaster Sound, Parry found clear water where Ross had marked Croker's Mountains on his charts. As Parry passed geographical features which appeared on no Admiralty charts, he mentally allocated names in honour of members of the royal family, Admiralty officials and others who had furthered his career: Barrow Strait, Prince Regent Inlet, Melville Island and Radstock Bay.

Parry's deck watch; image © The Worshipful Company of Clockmakers; Clarissa Bruce/Science Museum Group.

On 4 September, after a relatively unchallenging voyage, *Hecla* and *Griper* crossed 110°W. Parry, after double-checking his timepieces and longitudinal readings, duly advised his men they had earned a share of a Parliamentary award of £5,000. Other rewards appeared to be within their grasp, but Parry's hopes were soon dashed after his westward progress was blocked by huge quantities of ice.

In 1820 Parry was promoted to commander, elected a Fellow of the Royal Society and commissioned to lead another expedition. The following year, Parry was ordered to sail to Hudson's Bay, from where it was hoped he might find a more southerly west-bound route which bypassed ice in Barrow Strait. The theory, however, proved different to

'Cutting into Winter Island, Oct. 1821'; opp. p. 118, Parry, *Journal of a Second Voyage ... A North-West Passage*, author's collection.

the practice and this time Parry's ships, *Hecla* and *Fury*, barely reached 80°W before becoming frozen in. Notwithstanding the lack of progress along the Northwest Passage, Parry was promoted to captain and offered the post of acting Hydrographer of the Navy.

In 1824, after accepting another Arctic commission, Parry returned to Lancaster Sound and Barrow Strait on *Hecla* and *Fury* – which became ice-beset at around 90°W. During a long, dark winter in Prince Regent Inlet, Parry, with assistance from James Ross, Francis Crozier and other polar veterans, kept their men well fed, gainfully occupied and, thanks to entertainments including the 'Royal Arctic Theatre', in reasonably good spirits. Parry expected his ships to float free in early summer, but in July 1825, after *Fury* was crushed by ice, he abandoned her, left surplus rations and equipment on the beach and ordered all his men to join *Hecla*, so they could make their escape before the ice closed in again.

When Parry returned to London he was, as naval rules required, court-martialled for abandoning *Fury*. In light of the circumstances, he escaped censure and was confirmed in the post of Hydrographer to the Navy. Although Parry's attempts to traverse the Northwest Passage had suffered from the law of diminishing returns, he was offered command of an expedition to Spitsbergen, from where he, James Ross and others would attempt to outdo Buchan and Constantine Phipps and reach the North Pole and, should it exist, the Open Polar Sea.

When Parry left in April 1827, the Admiralty provided him with a Parkinson & Frodsham deck watch, other timepieces and navigational equipment and, for crossing expanses of sea ice, two custom-designed 20ft sledge boats, fitted with runners and detachable wheels.[2] In northern Norway, Perry and his men received skiing tuition and were shown how to harness and control the reindeer which would pull the sledge boats. But after they reached Spitsbergen and got under way, Parry's men resorted to pulling the sledge boats themselves. As the midsummer sun baked down on them, the exhausted men became sunburned, snow-blind and, after attempting to haul sledge boats over irregular ice floes floating on 2ft-deep water, increasingly despairing.

Although Parry had brought special twenty-four-hour chronometers to use during high summer, he and others regularly lost track of the time and distance travelled – calculations which became increasingly difficult after the ice over which they were travelling began to carry them south almost as fast as they could march north. In late July, their carefully calculated rations proved increasingly inadequate for the energy the men were expending. After Parry and his men passed 82°N he called a halt, knowing that while they were well short of 89°N (and a £5,000 reward), they were ahead of Phipps's and the Scoresbys' Farthest North records.

When Parry returned to London, he decided that as he was now married and had suffered badly from painful rheumatism during this expedition, he would draw his Arctic career to a close. The North Pole and Northwest Passage had eluded him, but he had established a new Farthest North and he had not only reached the Admiralty's Northwest Passage target latitude of 110°W but continued, against the odds, to what his deck watch, chronometers and other instruments showed him to be 113° 46' 43.5"W.

POLAR POSTSCRIPT: Parry resigned from his Admiralty post in mid-1829 when, partly for health reasons, he and his wife moved to the warmer climes of New South Wales, Australia. He later returned to the Admiralty and was appointed to a new post in which he oversaw the introduction of steamships into the navy.

8

A Canister of Meat

This unopened metal canister contains preserved meat and was produced for the Admiralty by Donkin, Hall & Gamble for Edward Parry's 1824–25 Arctic expedition. John Ross retrieved the canister and other provisions abandoned by Parry on Fury Beach, off Prince Regent Inlet, while on his own 1829–33 expedition, during which he hoped, as Parry had, to traverse the Northwest Passage.

John Ross presented the canister to his friend, John Johnstone, MP for Scarborough, after returning from his expedition. Johnstone presented it to the Scarborough Philosophical Society, who displayed it in their new Rotunda Museum, where it remains part of the collection today.[1]

In 1828, following Edward Parry's retirement from active service, John Ross submitted a proposal for an Arctic expedition to Admiralty officials. After his proposal was rejected, Ross persuaded his long-standing and philanthropic friend Felix Booth (director of a family gin-distilling business) to finance the purchase of *Victory*, an 85-ton paddle steamer with an auxiliary steam engine.[2] Ross also persuaded his nephew James, with whom he had not sailed since 1818, to serve as second-in-command.

The Rosses left London in May 1829 and three months later, after passing through Lancaster Sound and Barrow Strait, headed south down Prince Regent Inlet, where they hoped to find a navigable, ice-free westbound channel. After James Ross's readings suggested they were near the North Magnetic Pole, they dropped anchor off Fury Beach, where Parry had abandoned his ship and large quantities of canisters of meat and soup and other provisions – which John Ross used to supplement his own winter rations. As *Victory* continued south along Prince Regent Inlet, John Ross named an uncharted

Tin canister from Fury Beach; image © and courtesy of Scarborough Museums Trust.

extension of North Somerset 'Boothia Felix' in honour of his generous friend.[3] Their next stop was a sheltered inlet, which Ross named Felix Harbour and where the men prepared *Victory* for overwintering.

In early January 1830, a group of some thirty Inuit came to inspect *Victory*. As James Ross was by now familiar with Inuit dialect, friendly relations were established. During winter months the Inuit showed the visiting 'kabloonas' (as they would refer to non-Inuit) where to find salmon and how to make building blocks from snow, sledges from 'planks' of frozen salmon and garments from animal pelts. On learning that the Rosses were heading west, they indicated that as there was no westbound strait south of the Rosses' winter quarters, their visitors would need to return north before continuing west.[4]

'Somerset House on Fury Beach'; vintage print after John Ross, author's collection.

In mid-May, James Ross and a dog-sledging party headed west across Boothia Felix and crossed a frozen channel to what he assumed was an outcrop of the north Canadian coast. Following the coast of what he provisionally named 'King William's Land', he reached Cape Felix, from where he saw huge expanses of ice stretching north and west. Turning south-west, he reached Victory Point, where he and his companions built a 6ft-high cairn and deposited a canister with a report on their progress to date. Ross was now beyond 98°W and realised that he was only 200 miles east of Point Turnagain, the easternmost point Franklin had reached during his 1818–21 overland expedition.

After Ross's party returned to Felix Harbour, *Victory* was prepared for departure, but shortly after she left harbour she became frozen in again, leaving the Rosses with no alternative but to prepare for a second winter. No Inuit came to visit *Victory* during the winter of 1830–31, but the men remained scurvy-free and fit for the spring sledging season.

In mid-May, the Rosses, each leading a sledge, set out with ten expedition members, two Inuit guides and three weeks' provisions. As provisions ran low, John Ross's party returned to *Victory* to restock, leaving James Ross's party to explore further. Ross had, during Parry's expeditions, become interested in magnetism so he headed towards what, according to his readings, was the area of the North Magnetic Pole.[5]

'Snow cottages of the Boothians [Inuit]'; vintage print after John Ross, author's collection.

On 1 June, on the west coast of Boothia, his compass stopped working and the horizontal needle on his dip compass tilted to near-vertical. After rechecking his readings, Ross raised the Union Jack at around 70° 5'N 96° 46'W and told his men that they were now, more or less, at the constantly drifting North Magnetic Pole. Ross looked forward to sharing his findings with his friend Edward Sabine and other magnetism experts, but shortly after *Victory* was released from the ice she ran aground and broke her rudder, leaving the Rosses and their shipmates to face another winter.

By spring 1832, one expedition member had died, others were suffering from scurvy and John Ross's old war wounds were reopening. Rather than contemplate the possibility of another winter, the men began relaying auxiliary boats, sledges and provisions north towards Barrow Strait. On 29 May, John Ross dispensed tots of Booth's gin and formally announced that he was abandoning *Victory*. By early July, everyone was at Fury Beach, where men ate their fill from the dwindling piles of food canisters, then began building accommodation quarters – grandly named 'Somerset House' – from *Fury*'s timbers and Inuit-style blocks of snow.[6]

In early August, as the sea ice retreated, they loaded auxiliary boats with provisions for six weeks, then headed for Cape Seppings, just south of the entrance to Prince Regent Inlet. When Barrow Strait remained impassable, they deposited boats in Batty Bay and returned to Fury Beach to overwinter.[7] Although 'Somerset House' was reasonably spacious and warmed by portable heaters, the men were now subsisting on half-rations of eight-year-old canned provisions, so even a Christmas dinner of roast Arctic fox did little to boost morale.

In early 1833, carpenter Chimham Thomas died from scurvy which, together with the debilitated state of John Ross and others, suggested their remaining lime juice had lost its anti-scorbutic properties. As wildlife returned after the winter, however, regular meals of fresh meat soon improved the men's health and stamina – so much so that by mid-July, everyone was back at Batty Bay, bundling up cans and other rations in readiness for another attempt to reach Lancaster Sound.

By late August, the Rosses and their exhausted men had battled through gales and, by alternately rowing and sailing, reached an inlet off Lancaster Sound. Everyone collapsed on the beach exhausted, but at 4 a.m. on 26 August the lookout roused James Ross and pointed out a ship's sail on the horizon. They relaunched their boats, but the ship suddenly veered away.

Four hours later, another ship appeared and, as the Rosses and their men signalled, they saw a tender being lowered. As the tender came within hailing distance, John Ross asked its skipper the name of the ship. Ross, rarely at loss for words, was dumbstruck when he was told that it was '*Isabella* of Hull, once commanded by Captain Ross'. When Ross said that he was Captain Ross, his rescuers explained patiently that Captain Ross had gone to the Arctic but not returned. But by the time the Rosses and their men reached *Isabella*, confusion was dispelled, and whaling master Richard Humphreys and his crew gave the Arctic castaways a hearty welcome.

Back in London, the Rosses were welcomed as heroes and learned that George Back had travelled to Canada in hopes of relieving the Rosses by reaching them via the Great Fish River.[8] John and James Ross were both presented to the king, but it was John Ross who, as expedition leader, received most of the accolades. While James wrote up his magnetic observations and other reports, John interspersed expedition work with visits to friends, whom he regularly regaled with his travellers' tales and, on occasion, presented them with a still-unopened canister of food from Fury Beach.

POLAR POSTSCRIPT: The contents of the cans left at Fury Beach appear to have remained edible for long periods. Ross presented a can to a friend's son who, when he opened and ate from the can in 1869, pronounced the contents 'quite good' and apparently suffered no ill effects.[9]

9

James Ross's Career-Defining Portrait

This arresting portrait of James Ross by John R. Wildman was exhibited in the Society of British Artists' 1834 exhibition. It appeared at No. 1 in the catalogue and was captioned 'Commander James Clark Ross, R.N., etc., Discoverer of the North Magnetic Pole'.

The three-quarter-length oil portrait (*c.*170cm by 140cm in frame) is in the collection of the Royal Museums Greenwich/National Maritime Museum (ref. BHC2981).

In early 1834, shortly after returning from his sixth Arctic expedition, James Ross wrote to the First Lord of the Admiralty offering to command an expedition in search of a Northwest Passage.[1] When his proposal was declined, Ross knew he would soon return to half pay, with little prospect of an alternative commission.

During his early years in the navy, Ross was often referred to as 'John Ross's nephew'. He had become his own man when serving under Parry, so when he had signed up for his uncle's recent Arctic expedition, he had made a point of doing so under his full name, James Clark Ross.[2] On a one-ship expedition, however, James Ross could only claim to be second-in-command, but after his uncle was awarded the Royal Geographical Society's annual Founder's Medal, some suggested that James should have received a share of the 50 guineas award and his name should have appeared on the citation.[3]

It was a long-standing naval custom that officers returning from long-distance expeditions sat for portraits when they returned to Britain.[4] In January 1934, a new portrait of John Ross by Royal Academician James Green was unveiled at a private *conversazione,* prior to it being exhibited at

Portrait of James Clark Ross, John Wildman, 1834 (ref. BHC2981); image © National Maritime Museum, Greenwich, London.

the Royal Academy exhibition in May.[5] In the portrait the senior Ross, clad in bearskin-swathed dress uniform, looked impressive, but when the two Rosses attended a Raleigh Club meeting later that month, a subsequent report referred to James as 'the most distinguished of guests', describing him as 'a man of intelligence and resolution', whose face was 'remarkable for sharpness and vivacity'.[6]

The Admiralty had rejected James Ross's proposal for his own expedition, but he was already, by common consent, 'the discoverer of the North Magnetic Pole', so when he sat for portraitist John Wildman,

Portrait of John Ross, (attr.) Henry Hawkins, *c.*1834 (ref. BHC2983); image © National Maritime Museum, Greenwich, London.

he wore bearskin-draped dress uniform and he asked Wildman to include in the foreground his magnetic dip circle and, in the dark sky background, a gleaming North Pole Star.

When the Society of British Artists' spring exhibition opened in late March, one reviewer commented that Wildman's painting merited its No. 1 ranking in the catalogue, while another praised the portrait of James Ross as 'spirited and impressive' and 'full of animation'.[7] Green's portrait of James's uncle could not be publicly exhibited until the Royal Academy's May exhibition, but Henry Hawkins, a long-standing member of Society of British Artists, had submitted a portrait of him in which the explorer, clad in a sealskin coat, was portrayed in front of what looked like a painted panorama panel or theatrical backdrop of Felix Harbour – a composition one reviewer suggested was 'not well conceived or arranged'.[8]

By May, Green's and other portraits of John Ross could be seen at the Royal Academy exhibition. Lithographs of portraits of him were also widely available, while representations of both Rosses could also be seen in Robert Burford and Henry Selous's new panorama of Boothia, which continued to draw crowds to the Leicester Square Rotunda.[9]

John Ross had closely supervised the production of the Boothia panorama, as he did the development of an Arctic extravaganza which would soon open at Vauxhall Gardens. The 60,000 square feet of painted canvasses would form the backdrop to models of fur-costumed Inuit and polar bears, whales (which spouted water), icebergs over 70ft high, large-scale models of *Victory* and other vessels and, emerging heroically from a massive thunderstorm, a model of John Ross, in all his glory.[10]

John Ross, following his years in obscurity, clearly relished being the toast of the town, but James was more interested in completing his expedition reports and wooing 17-year-old Anne Coulman, the daughter of a friend of his sister. Although the handsome 34-year-old soon won Anne's heart, her father pointed out that she was too young to marry anyone, let alone someone who spent most of his time away from home.[11]

Panorama panel of rescue by *Isabella* (based on Ross's drawings), now in North West Castle Hotel, Stranraer (Ross's former home); image © A. Strathie, with permission of Bespoke Hotels.

Meanwhile, John, now in his late fifties, was courting 22-year-old Mary Jones, the sister of his friend Dr John Rymer-Jones.[12] By the end of the year, Sir John Ross (as he now was) had married Mary and was spending more time at his home in Stranraer, which he planned to extend and enhance, including with a large-scale depiction and mementos of his celebrated expedition.[13]

The year also ended well for James, who had been promoted to captain, and, in recognition of his work on magnetism, elected a Fellow of the Royal Society. The portrait which portrayed him as 'Discoverer of the North Magnetic Pole' was much admired (including by John Franklin's second wife, Jane), but it was Ross's readings and reports that earned him his next commission, as a participant in a major magnetic survey of the British Isles, organised by the Admiralty's adviser on magnetism, Edward Sabine.

POLAR POSTSCRIPT: James Ross and Edward Sabine first met when serving on John Ross's 1818 expedition on *Isabella*. They then served together on Parry's 1819–20 expedition. James was only 18 when they first met, but Sabine, after noticing his interest in science, encouraged his research on magnetism. Ross worked on Sabine's survey until 1839, apart from a short period in late 1835 when Ross led a relief expedition to assist ice-beset whalers trapped in Baffin Bay. During the expedition, Ross's shipmaster was Richard Humphreys (who had recently rescued Ross on *Isabella*) and his No. 2 was Francis Crozier, with whom Ross had served on three previous Arctic expeditions.

10

Rossbank Magnetic Observatory, Hobart

Lady Jane Franklin commissioned this painting of Rossbank Magnetic Observatory from British artist Thomas Bock, one of Hobart's leading artists.[1] The observatory was built in 1839 as one of a chain established by James Ross and Francis Crozier during their 1829–33 circumnavigation. The figures in the foreground (left to right) are John Franklin, James Ross and Francis Crozier; in the background is Franklin's nephew, Lieutenant Henry Kay, who worked at the observatory. Jane Franklin knew Bock, for whom she had sat for a portrait the previous year.

The painting is in the collection of the Tasmanian Museum and Art Gallery, in Hobart, where it was transferred in 1948 from the Scott Polar Research Institute, Cambridge, England (to whom it was bequeathed by the Lefroy sisters, John Franklin's great-nieces). A version of the painting was presented to Edward Sabine, the Admiralty's adviser on magnetism. The observatory was demolished in the 1850s.

In mid-August 1840, HMS *Erebus* and *Terror* docked in Hobart. The ships' captains, James Ross and Francis Crozier, had left London almost a year previously, but they had no time to relax because they only had two weeks to erect a fully operational magnetic observatory to standards laid down by the Admiralty's magnetism adviser, Edward Sabine. Once built, the Hobart observatory would be the most southerly in a chain of observatories stretching south from Canada. On their way south, Ross and Crozier had also transported Royal Artillery teams led by Henry Lefroy and Frederick Eardley-Wilmot, who should by now have established observatories on St Helena and the Cape of Good Hope.

Rossbank Observatory, Hobarton, Thomas Bock (1790–1855); in collection of and image © Tasmanian Museum and Art Gallery, Hobart (oil painting, ref. AG241).

When Ross and Crozier landed, they were warmly greeted by John Franklin, who had been offered no naval commissions since his last Arctic expedition and had accepted the post of Lieutenant Governor of Van Diemen's Land (Tasmania) two years ago. Franklin had already identified a suitable site for the observatory in the grounds of his residence and arranged for about 200 deportees to assist with construction work. This was becoming a Franklin family affair as Franklin's nephew, Lieutenant Henry Kay, had travelled south with Ross and Crozier and would stay on to operate the observatory.

The observatory was up and running in time for August's magnetic reading 'term days' when readings from all over the world were collated and compared. Before continuing their circumnavigation, Ross and Crozier trained Kay and Franklin so they could run the observatory while they headed south to take more magnetic readings and, all being well, locate the South Magnetic Pole.

Lady Jane Franklin by Thomas Bock (chalk on paper, c.1838); in collection of and image © and courtesy of Queen Victoria Museum and Art Gallery, Launceston, Tasmania.

While in Hobart, Ross received updates on the progress of Frenchman Jules Dumont d'Urville and American Charles Wilkes, who were, he knew, also trying to locate the South Magnetic Pole. To his relief, while both had apparently sighted land south of 66°S, neither was close to areas that Ross planned to explore.

To commemorate the establishment of the 'Rossbank' (as the observatory was named), Jane Franklin had commissioned a painting from artist Thomas Bock, who had arrived in Hobart as a deportee, but after being pardoned had, with support from other enlightened patrons, re-established his career as an artist.[2] Jane Franklin wanted Bock to include Franklin, Kay, Ross and Crozier in the painting, but as Ross wanted to make the most of the Antarctic summer, Crozier arranged for *Terror*'s sailing master, John Davis (himself a talented artist), to provide Bock with a watercolour of the two captains in front of the observatory.[3]

Erebus and *Terror* left Hobart on 12 November and headed for the sub-Antarctic Auckland Islands, where Ross found two hand-painted signboards that had been left by d'Urville and Wilkes, indicating that they had both watered ship there in March. Their messages suggested that d'Urville had discovered '*la Terre Adélie*' and the position of the South Magnetic Pole, and that Wilkes was simply undertaking an 'exploring cruize along the Antarctic circle'.[4]

On New Year's Day 1841, after *Erebus* and *Terror* crossed the Antarctic Circle, Ross and Crozier issued celebratory double measures of rum, extra rations and additional winter clothing. Ten days later, at around 71°S, Ross was surprised to hear a cry of 'Land ahoy!' and to see, under the midnight sun, distant ranges of mountains, rising to around 10,000ft. Ross, who, like Parry, enjoyed naming uncharted geographical features, named the tallest mountain in sight Mount Sabine, for his magnetism mentor, and a headland Cape Adare, for one of his expedition supporters. After checking his coordinates, Ross realised that he was correct in estimating that the Magnetic Pole lay around 76°S 145°E – not on the coast, as he had hoped, but in the middle of a considerable land mass.

As *Erebus* and *Terror* continued south, Ross named uncharted geographical features for notable individuals and institutions, but on 17 January, the birthday of his sweetheart Anne Coulman, he named islands for her uncle and father and Cape Anne for her.[5]

As his ships passed Weddell's Farthest South of 74° 15', Ross issued more celebratory grog. As they approached 76°S, the latitude at which Ross reckoned the Magnetic Pole lay, Ross began looking for places to land. When he realised there was too much shore ice, he felt frustrated, but before long he was confronted by sights which took his mind, temporarily at least, off the Magnetic Pole.

POLAR POSTSCRIPT: Henry Kay remained in charge of the Rossbank Observatory until its closure in 1853. He was assisted for several years by Jane Franklin's nephew, Francis Simpkinson. Kay's architect brother, William Porden Kay, also came to work in Hobart, where he worked on the design and construction of a new governor's residence adjacent to the Rossbank site.

11

A Great Icy Barrier

The 'Great Icy Barrier', now known as the Ross Ice Shelf, was first seen by James Ross, Francis Crozier and their shipmates in January 1841. It rises to around 200ft and measures almost 500 miles by 600 miles.

Images of the 'Barrier' made by *Terror*'s second master, John Davis, show it as it was in 1841, since when, based on reports of later expeditions, it has receded. Davis made several images of the barrier, some of which were published in Ross's expedition report in 1847.[1]

On 28 January 1841, James Ross saw, due south of *Erebus* and *Terror*, a 12,400ft volcano, from which issued smoke and flames and, on the same land mass, a slightly smaller, apparently dormant volcano. As Ross and Crozier looked in wonder at the newly named Mount Erebus and Mount Terror, they noticed a huge ice cliff running eastwards from the foot of Mount Terror, which continued as far as the eye could see. The ice cliff rose to around 200ft and Ross reckoned it might be up to 1,000ft thick. Even from the crow's nest, it looked as if the ice shelf stretched south to the horizon, although Ross thought he could see a mountain range (which he charted as 'Parry Mountains') in the far distance.[2]

The barrier seemed to Ross to be as impenetrable as the white cliffs at Dover and although it appeared to have no fissures running through it, he kept *Erebus* and *Terror* a safe distance from it as they sailed parallel to it. When fog and snowstorms obscured the barrier from view, Ross and Crozier retreated from the barrier and from each other and kept track of each other's position by exchanging musket volleys. At 77°S 187°E, the barrier dipped sufficiently for Ross to see over its flat top, but dense pack ice between the ships and the barrier meant there was no chance of landing.[3]

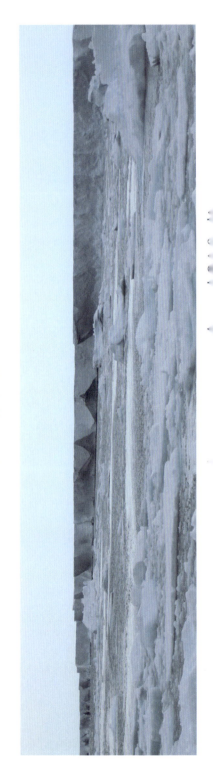

'Part of the South Pole Barrier ... [180ft high, 1,000ft thick, 450 miles long] ... February 2 1841', John Davis; from *A Voyage of Discovery and Research*, vol. I, opp. p. 233, author's collection.

Ross was tempted to continue east and try to locate the terminus of the barrier, but as he wanted to make an early start for the Magnetic Pole the following season, he reluctantly decided to return west. When they reached the volcanic island, they were only about 160 miles from the Magnetic Pole, but as newly named Cape Crozier and other potential landing sites proved inaccessible, Ross decided to leave the area. As *Erebus* and *Terror* headed north, Mount Erebus sent plumes of smoke and tongues of fire into the sky, which was filled with curtains of aurora more beautiful than any Ross and Crozier remembered seeing in the Arctic.

On the return voyage to Hobart, Ross followed coasts visited by d'Urville, Wilkes and the sealer John Bellany, rechecking any readings or charts he thought were inaccurate.[4] When they reached Hobart in early April, Ross and Crozier learned that Franklin's wife Jane and niece Sophia Cracoft were travelling in New Zealand. In May, as their second pleasant stay in Hobert drew to a close, Ross and Crozier decided to repay the Franklins' hospitality by organising a shipboard ball and dinner on 1 June, the tenth anniversary of Ross reaching the North Magnetic Pole.

Jane Franklin was still in New Zealand, but Franklin, his daughter Eleanor and niece Sophia (to whom Crozier had taken a liking) were at home. The evening was a great success and the Franklins and their fellow guests enjoyed champagne, a lavish dinner and dancing until dawn on the lantern-bedecked ships. Hobart belles vied to dance with Ross and Crozier, but Ross remained faithful to Anne Coulman, and Crozier had eyes for none but Sophia (or 'Sophy') Cracroft – who unfortunately seemed more interested in Ross.

On 7 July, *Erebus* and *Terror* were cheered away from Hobart. After several months of magnetic work in Australia, New Zealand and on sub-Antarctic Islands, *Erebus* and *Terror* returned south. On 1 January 1842, after crossing the Antarctic Circle, the men secured them both to a large, flat-topped iceberg where everyone enjoyed a double celebration, complete with entertainments and dancing led by Crozier and 'Miss Ross' – a character Ross had created for Parry's 'Royal Arctic Theatre'.

Their voyage south was plagued by blizzards, fog, storms, gales and dense pack ice, but they managed to continue soundings and collecting specimens of different species of penguins. On 22 February, they reached the Great Icy Barrier and, from roughly the longitude they had turned back in 1841, began following it eastwards. The following day they recorded, at 161° 27'W, a Farthest South just beyond 78°S. Ross wondered whether to return west and attempt another landing near the Magnetic Pole, but as temperatures dropped he decided to head north to the Falkland Islands.

In mid-March, through swirling snow, Ross saw a chain of icebergs bearing down on *Erebus* and *Terror*. He ordered his men to halt *Erebus* by reefing her sails, but as she slowed he saw *Terror* racing away from a 200ft iceberg and heading straight for *Erebus*. As the two ships' masts and riggings intertwined, they tossed and rolled helplessly in the waves until *Terror* suddenly sprang free and raced between two huge icebergs, leaving Ross and his men struggling to control their badly damaged ship.

As *Erebus* began banging against an iceberg, Ross had no choice but to use a complex 'stern-board' manoeuvre normally reserved for calm seas. But as *Erebus* pulled back from the iceberg, Ross saw another huge one approaching. This time, he ordered his men to wheel *Erebus* round and, after getting more wind in her sails, propel her between the next two icebergs so she could rejoin *Terror*.

After their narrow escape, Ross and Crozier were glad to reach the Falkland Islands after four months at sea. When they docked in Port Louis, *Erebus* and *Terror* were thoroughly overhauled, while Ross, Crozier and a working party erected another observatory and others investigated the seashore and wiry tussock grass moors.[5] At the South Shetland Islands, they charted Erebus and Terror Gulf, Snow Hill (so-called for its appearance) and Seymour Island. While passing Joinville Island, Ross named a tower-shaped rock D'Urville Monument in memory of the French explorer who had, Ross had learned, recently died with his wife Adèle (for whom d'Urville had named a species of penguin) in a train crash in Paris.

Thick pack ice in the Weddell Sea prevented Ross from taking magnetic readings, but they visited the magnetic observatories at the Cape of Good Hope and at St Helena, where Ross learned that Henry Lefroy had, with Sabine's agreement, transferred to the Toronto observatory.

On 5 September 1843, *Erebus* and *Terror* returned to Britain after an absence of over four years.[6] Sabine and other Admiralty colleagues were pleased with Ross and Crozier's magnetic work. Ross was aware, however, that the public had limited interest in scientific work, so when preparing his expedition report, *A Voyage of Discovery and Research*, for publication, he included accounts and illustrations of the perilous encounter with the chain of icebergs and of the Great Icy Barrier – which he described as 'a mighty and wonderful object', something that he considered, despite his extensive Arctic travels, far beyond anything he could have previously imagined.

POLAR POSTSCRIPT: The name Ross Sea first appeared on charts in the early 1900s, but the Great Icy Barrier was only renamed the Ross Ice Shelf in the 1950s.

12

Francis Crozier's Penguin

This specimen of a 'great penguin' was brought back from Antarctica by Francis Crozier from his 1839–43 circumnavigation on HMS *Terror*. Most of the expedition's zoological specimens were deposited with the British Museum in London, but Crozier presented this and other specimens to the Belfast Natural History and Philosophical Society, who ran the nearest natural history museum to his home town of Banbridge, County Down.

The penguin specimen, now recognised as an emperor penguin, or *Aptenodytes forsteri*, is now in the collection of the Ulster Museum, Belfast.

In late 1843, Francis Crozier presented two pickled penguin skins and about 100 other specimens to the Belfast Natural History and Philosophical Society, which had a museum where Crozier hoped his collection might be displayed. The curator politely indicated that there was a shortage of space, but the society's president, William Thomson, enthused by a recent scientific voyage with the distinguished Scottish naturalist Edward Forbes, agreed to accept Crozier's gift.[1]

Crozier and his shipmates had, during their circumnavigation, seen tens of thousands of penguins on islands, seashores, ice floes and pack ice, and watched them swimming, 'porpoising' or diving in and out of the ocean. Penguins came in all sizes, ranging from the smaller Adélie and crested penguins to king penguins, which were much taller and up to three times as heavy as the smaller species.

As *Erebus* and *Terror* followed the Great Icy Barrier, naturalists Robert McCormick, Joseph Hooker and David Lyall saw what they thought were particularly large king penguins standing and lying on ice floes. When they examined their specimens, however, they discovered that while they

Emperor penguin specimen donated by Francis Crozier (ref. BELUM.Lg5110); image © and courtesy of National Museums Northern Ireland, Ulster Museum Collection.

were only about 10 per cent taller than king penguins they had seen elsewhere, some weighed almost 80lb and had orange-gold markings around their head and neck which were much more vivid than similar markings on king penguins.

These 'great penguins', as Crozier referred to them pending clarification and potential reclassification, also seemed calmer and less sociable than king penguins and would sometimes stand alone or in small groups, or process across the ice at a stately pace. But if men tried to capture them, they flopped forward onto their bellies and, using their feet and flippers, propelled themselves at impressive speeds across the ice. To avoid wasting parts of the birds that scientists might not need, expedition cooks experimented with the dark, fatty meat and devised a hearty soup which was so similar to hare soup that few could tell the difference. McCormick, Hooker and Lyall agreed the soup was delicious but they were more interested in establishing whether they had discovered a new

Crozier's and other existing large-sized specimens were compared with king penguins for size and colouring before emperor penguins were classified as a new species; images © A. Strathie.

species – something they hoped John Gray, the British Museum's Keeper of Zoology, would soon tell them.

Crozier had already learned a good deal about magnetism and natural science from Ross, Parry, Sabine and the three scientists with whom he had recently sailed. He was keen to learn more, but as Ross was now engaged to Anne Coulman, it seemed likely that Anne's father would hold Ross to his promise of not accepting further long-distance or potentially dangerous commissions. For Crozier, this suggested that over a decade of polar voyages with Ross, his mentor and closest friend, was coming to an end.

Following their circumnavigation, Ross had been knighted and presented with the Royal Geographical Society's Founder's Medal. Meanwhile, Crozier had been promoted to captain (albeit after thirty years' naval service) and elected to membership of the Royal Society and the Belfast Natural History and Philosophical Society. As to settling down and marrying, Crozier had remained fond of John Franklin's niece,

Sophy Cracroft. After leaving Hobart for the second time, Crozier had tried to put her out of his mind, but after the Franklins and she returned to England in summer 1844, he had suggested they meet again. Sophy had been polite, but after they had met a few times, Crozier realised that although she appeared to like him, she was not interested in marrying him and remaining at home while he travelled the world.

While Crozier awaited his next commission, John Gray, the British Museum's Keeper of Zoology, made known his decision regarding Crozier's and other specimens of so-called 'great penguins'. As Crozier and Ross had hoped and McCormick, Hooker and Lyall had suspected, the birds were a different species from *Aptenodytes patagonicus*, the Patagonian king penguin. Although they were also *Aptenodytes*, or 'featherless divers', Gray decided that, rather than being described for their location, they should be named *Aptenodytes forsteri* in honour of Johann Forster, who had identified several species of penguins while travelling with Cook and may, albeit unknowingly, have seen these larger birds on the ice.

Francis Crozier was gratified to have been involved in the identification of a new species and to have presented the Belfast Natural History and Philosophical Society with a specimen which members of his family and local friends might be able to see for themselves. He felt fortunate to have sailed with Ross for so many years and, particularly on this last voyage, to have seen so much of the world. Crozier was very fond of Anne Coulman and wished her and 'her darling James' every happiness – but as for his own career and marriage prospects, he would just need to wait and see what the future held in store.

POLAR POSTSCRIPT: The breeding cycle of the emperor penguin (as they became known) was not well understood until naturalist Edward Wilson and others visited their Cape Crozier breeding grounds during Robert Scott's *Discovery* and *Terra Nova* expeditions.

Part III

The Northwest Passage: The Search Continues

In late 1844, as 80-year-old John Barrow approached retirement after four decades as the Admiralty's Second Secretary, he began planning the navy's first expedition to the Northwest Passage for over a decade. When James Ross and Edward Parry declined the potential commission, the shortlist was reduced to John Franklin (recently returned from Hobart and keen for a commission), Francis Crozier (*Terror*'s most recent captain) and 31-year-old Lieutenant James Fitzjames (Barrow's current protégé).[1]

In February 1845, shortly after Barrow went into retirement, the Admiralty announced that Franklin would command the expedition and *Erebus*, Crozier would captain *Terror* and Fitzjames would serve as *Erebus*'s second-in-command, with additional responsibility for recruitment and magnetic observations.[2] Franklin's instructions were to enter Lancaster Sound, continue along Barrow Strait past Prince Regent Inlet and North Somerset, then search for a navigable channel leading south-west towards Canada's northern coast. Should his ships become blocked by ice around Prince Regent Inlet, he would head north through Wellington Channel and try to find a more northerly route to the Bering Strait.

By the time Franklin left, Barrow was working on a sequel to his 1818 publication, *A Chronological History of Voyages into the Arctic Regions*, on the assumption that, should ice conditions prove favourable, he could complete his book – and his career – with a report on the successful navigation of the Northwest Passage.[3]

13

A Daguerreotype

This daguerreotype of 25-year-old Harry Goodsir, the assistant surgeon and naturalist on HMS *Erebus*, was one of a set of fourteen commissioned by Jane Franklin in May 1845 from daguerreotype patent-holder Richard Beard.[1] Of the fourteen daguerreotypes, thirteen are of *Erebus* officers and other senior personnel; the other is of Francis Crozier, captain of HMS *Terror*.

The daguerreotype is sixth-plate size (2¾in by 3¼in) and is one of twelve held by the Scott Polar Research Institute (SPRI) (ref. N:589/6). The set of daguerreotypes came to SPRI through John Franklin's great-nieces, the Lefroy sisters, to whom they were left by their aunt, Sophy Cracroft, the main beneficiary of Jane Franklin's will.[2] It is uncertain when (assuming the twelve formed part of a full set) the daguerreotypes of Francis Crozier and *Erebus*'s ship's mate, Robert Sargent, were separated from the others.

On 18 May 1845, as Harry Goodsir and other members of John Franklin's Arctic expedition prepared to leave London, a camera operator representing daguerreotypist Richard Beard set up a temporary on-deck studio on *Erebus*. Harry had been photographed by a family friend, Dr John Adamson, a few years previously, so decided to sit in a position where he could support his head with his hand and hold his pose for as long as required while the plate was exposed.[3] The operator explained that, although Lady Franklin would receive a full set of daguerreotypes, Harry or his relatives could order copies of his image or others from Beard's King William Street studio – something Harry remembered to tell his sister Jane when next writing to her.[4]

Since arriving in London in March, Harry had been preparing for the expedition and working on scientific papers he had promised to

Daguerreotype of Harry Goodsir (Richard Beard's studio), in original case; image © and with permission of Scott Polar Research Institute, University of Cambridge.

send to his brother John in Edinburgh. He found it difficult, however, to decline invitations from his friend Edward Forbes (with whom he had studied and roomed in Edinburgh) to attend meetings of learned societies and Forbes's 'Red Lions' scientists' dining club.[5] It was, as Harry told his sister, 'a new mode of life', which left little time for purchasing the sealskin greatcoat, initialled silver cutlery, deep-sea dredging nets and other items he needed for the voyage. As the money Harry brought from Scotland soon ran out, he was grateful when new Admiralty rules resulted in his receiving a pay increase for serving as *Erebus*'s naturalist as well as her assistant surgeon.[6]

While Beard's representative was on board, Jane Franklin visited the ship and wished Harry and his companions well. She seemed, Harry thought,

A sketch of Harry Goodsir by Edward Forbes, drawn shortly before Goodsir left London (ID: GPS-Goodsir-1); image © and courtesy of University of St Andrews Libraries and Museums.

Photograph of Harry Goodsir, taken in 1842 by Dr John Adamson (ID: ALB-8-90); image © and courtesy of University of St Andrews Libraries and Museums.

rather 'overcome' and concerned for her 59-year-old husband, who was suffering from a heavy cold. Before long, however, Franklin seemed in excellent spirits and, after enquiring about Harry's natural history work, suggested he use the large table in the spacious captain's cabin to write up his reports – an offer Harry soon took up as, in addition to his work for his brother, he was now also writing up zoological reports which Crozier wanted to send back from Baffin Bay with the expedition's supply ships.

On 31 May, *Erebus* and *Terror* reached Stromness, where Hudson's Bay Company and other ships regularly reprovisioned and rewatered at Login's Well. The company had offices in Stromness and regularly recruited Orcadians, including Dr John Rae, whose mother and sisters entertained Franklin and several officers during their stay.[7]

As *Erebus* and *Terror* sailed north to Disko Bay in Greenland, Harry worked long hours, including under the midnight sun. Although Harry's immediate superior, Dr Stephen Stanley, showed little interest in his natural science work, Franklin suggested Harry send copies of his reports to Dr John Richardson, who had accompanied Franklin on two overland Arctic expeditions and had, until the recent death of his wife, been related to Franklin by marriage.[8]

At Whale Fish Islands in Baffin Bay, Harry dredged up sea creatures from 300 fathoms, the limit at which his friend Edward Forbes believed marine life could exist.[9] Although Harry's medical knowledge had not been tested so far, men began complaining about bites from mosquitos which, like those of Scottish midges, left victims with red, itchy blotches and, in severe cases, badly swollen faces. When Harry went ashore to visit a local community, he was asked to help a man suffering from consumption, but the case was so advanced that Harry could do very little for him.

Although Harry still had scientific reports to finish for his brother and Crozier, he accepted Franklin's invitation to join him and a few others in an auxiliary boat for a tour of some huge icebergs. Harry was fascinated by one which had bright blue veins running through it and (with Franklin's permission) clambered onto it so he could inspect the veins at close quarters and collect ice specimens for later analysis.

As the expedition's last support vessel prepared to return south, Harry completed family letters and handed Crozier the scientific reports he would be sending south. Harry had not managed to finish the reports for his brother but bundled them up with specimens and a note of apology. Everyone was, it seemed, trying to tie up loose ends, including ice master James Reid, who was helping Harry with dredging and was concerned that his wife had made no mention of receiving the copy of the daguerreotype of himself that Beard's operator had promised to send her free of charge. Harry, who assumed his sister had ordered copies or passed Beard's address on, was currently more concerned about telling his father and brother John about the icebergs and other fascinating things he already seen before even entering the Northwest Passage.[10]

POLAR POSTSCRIPT: In letters from Greenland, Franklin suggests that Goodsir was a great 'favourite' on *Erebus*. No further copies of Goodsir's daguerreotype have been traced, although a copy of it may have been exhibited in Edinburgh in 1895.[11] Beard is recorded as having supplied daguerreotype equipment to the expedition and it is suggested that Goodsir may have been entrusted with the camera, but no images have yet been found.[12]

14

A Rock at Port Leopold

The rock, carved with 'E.I.' and '1849', stands on the foreshore of Port Leopold, near the junction of Barrow Strait and Prince Regent Inlet (74°N 90°W). The rock was carved during James Ross's 1848–49 relief expedition on *Enterprise* and *Investigator* when he left supplies and equipment at Port Leopold for John Franklin and members of his expedition.

The pyramid-shaped rock (around 3ft each way) stands on a prominent position on the foreshore. Like a similarly shaped rock that was left during Edward Parry's 1819–20 expedition (of which Ross was a member), it appears to be intended as a marker to indicate which ships had been there or signal the presence of provisions or other items that might be of assistance to others.

In early 1847, both John Ross and John Richardson became concerned about the lack of news from their good friend John Franklin.[1] By early 1848, following energetic lobbying by Jane Franklin, Admiralty officials were preparing for a three-pronged search of the Northwest Passage. HMS *Herald* (already in the Pacific) and *Plover* would search west to east from the Bering Strait, while Richardson led an overland party to the west of the archipelago and James Ross would command *Enterprise* and *Investigator* during an east-to-west search from Lancaster Sound.

Although Ross had felt bound by his earlier promise not to accept a long-distance or potentially dangerous commission, his wife Anne was very fond of Frank (as she called Crozier, who had been best man at their wedding) and was all in favour of her husband joining the search for him, Franklin and their shipmates.[2] In early 1848, the couple wrote letters to Crozier, wishing him a 'happy and triumphant return', then delivered

Rock with inscription on foreshore of Port Leopold (August 2023). The inscription signals the presence of *Enterprise* and *Investigator* in 1849 and of provisions and equipment left for Franklin and his men; image © A. Strathie.

them to the captain of *Plover*, which was about to leave Britain for the Bering Strait.[3]

In May 1848, Ross bade his beloved Anne farewell and sailed from Britain with trusted polar hands including Edward Bird, Robert McClure and Thomas Abernethy, and Lieutenants Leopold McClintock (well travelled and highly recommended) and William Browne (a talented draughtsman and artist). Ross had with him additional provisions to leave for Franklin and his men, letters and gifts from Jane and Eleanor Franklin and, at Jane Franklin's request, leaflets she wanted Ross to distribute to whalers, so they would be aware of rewards of up to £2,000 that she was offering to those who assisted her husband.[4]

When Ross reached Greenland he learned that, as winter 1846–47 had been particularly severe, Baffin Bay was still full of ice. It took him until

W.H. Browne, 'Noon in Winter' (Port Leopold); image courtesy of New York Public Library.

August to reach Lancaster Sound, where he made a landing at a cairn, where he thought (and hoped) Franklin or Crozier might have left a note. During a second landing, men built a second cairn where Ross deposited a cylinder with messages for his friends.

As the weather in Barrow Strait turned stormy and foggy, Ross ordered his men to discharge guns, fire rockets and let off flares in hopes of alerting Franklin's men or any Inuit nearby – who might have seen Franklin's ships – to their presence. There was no response, so Ross headed for Port Leopold, a sheltered bay near the junction of Barrow Strait and Prince Regent Inlet, where he wanted to deposit provisions for Franklin. Ross planned to leave the following day, but after huge quantities of ice swept into the bay, *Enterprise* and *Investigator* became completely hemmed in, leaving Ross no alternative but to ask his men to erect tarpaulins over the ships and prepare to overwinter.

As little had been achieved so far, Ross struggled to keep his men's morale up during the winter and by late April everyone was looking forward to the spring sledging campaign. In early May, sledge parties radiated out from Port Leopold, searching, depositing rations and building cairns at Cape Seppings and other prominent locations.

W.H. Browne, 'The Bivouac' (Cape Seppings); image courtesy of New York Public Library.

In mid-May, Ross, McClintock and twelve others embarked on a thorough search of North Somerset's north and west coasts. For six weeks, they searched coves and inlets and its uncharted western coastline, from where they could see a large, ice-covered body of water and more land.[5] Ross was keen to continue south to Boothia and check the current location of the North Magnetic Pole, but as the men began to struggle, he left a message in a newly built cairn on Cape Coulman and led the way back to Port Leopold.

When Ross, McClintock and their men returned to the ships in late June, they had travelled some 400 miles in gruelling conditions and many were suffering from scurvy.[6] By August, some were sufficiently recovered to join ice-sawing parties, help build emergency shelters for Franklin's men or work on the foreshore, where Ross wanted to leave one of *Investigator*'s steam launches and a large rock carved with his ships' initials and the date, which should let Franklin and Crozier know there were relief supplies to hand.

On 28 August, after almost a year in Port Leopold, *Enterprise* and *Investigator* finally escaped from the ice into Prince Regent Inlet and from there into Barrow Strait. Ross planned to update his last season's messages and check for new ones, before continuing to Melville Island, but the ice was too thick. While Ross was considering his options, a sudden gale swept tons of loose ice against the ships. Before long, *Enterprise* and *Investigator* were frozen fast into a 50-mile-wide block of ice, which was soon being propelled by westerly winds eastwards along Barrow Strait at a rate of about 10 miles a day.

Three weeks later, the huge floe and its involuntary passengers emerged into Baffin Bay – where the floe shattered and *Enterprise* and *Investigator* suddenly floated free. Although Ross had planned to overwinter and continue to search westwards in spring, the combination of his men's poor health and westerly winds which currently made re-entering Lancaster Sound nigh on impossible, led him to decide to return to Britain.

Although Admiralty officials and Jane Franklin made it clear they expected Ross to return and continue his search, there was now no shortage of volunteers following the government's announcement of rewards of up to £20,000 for relieving or 'ascertaining the fate' of Franklin's expedition. After Jane Franklin accepted that the exhausted and dispirited Ross would not go back, Richard Collinson and Ross's *Enterprise* first lieutenant, Robert McClure, took command of *Enterprise* and *Investigator* and headed to the Pacific Ocean and the Bering Strait.

Ross regarded his search for his best friend and his companions as a failure. He filed a report with the Admiralty but declined to produce the now customary illustrated expedition narrative. All he could hope was that Crozier, Franklin or some of their shipmates might find one of his messages or food depots, or the carved rock which would signal the presence of the steam cutter and provisions at Port Leopold.

POLAR POSTSCRIPT: In January 1850, Robert Burford (now owner of the Leicester Square Rotunda) showed a new panorama, 'Summer and Winter Views of the Polar Regions', which he and artist Henry Selous had created based on Lieutenant Browne's drawings. The Admiralty also published *Ten Coloured Views Taken During the Arctic Expedition of Her Majesty's Ships 'Enterprise' and 'Investigator'*, which contained some of Browne's finest works and a short expedition narrative.

15

John Rae's Octant

This octant belonged to Orcadian Dr John Rae who, after ten years as a Hudson's Bay Company medical officer, was commissioned to survey uncharted areas of the Arctic archipelago and adjacent coastline. Octants and sextants both measure latitudes by reference to celestial bodies. Octants are often made from wood and are robust and easier to use than sextants, so they were suitable for those, like Rae, who travelled long distances in remote areas.

Rae's octant, which measures around 17in by 19in, is in the collection (ref. E006) of Stromness Museum, Orkney. This is an independent museum run by the Orkney Natural History Society, which was founded in 1837.

In late October 1847, Dr John Rae arrived at HBC's London offices and reported on his recent surveying expedition to the Gulf of Boothia. Rae, who had joined HBC as a medical officer, regularly travelled long distances overland and by river in the company of Inuit guides and French-Canadian *voyageurs*. Sir George Simpson, head of HBC's North American operations, commissioned Rae to survey areas which neither British Naval officers nor HBC surveyors, Peter Dease and Thomas Simpson had charted. He hoped that Rae might also identify a continuous Northwest Passage route from Baffin Bay to the Bering Strait.

Before Rae had embarked on his surveying expedition, he was trained on the use of octants, dip circles and other navigational instruments by army officer Henry Lefroy, head of Toronto's magnetic observatory.[1] To carry out the survey, Rae and his small party had dragged boats overland from Hudson's Bay to the Gulf of Boothia and, over two seasons, charted

Octant belonging to John Rae; image © and courtesy of Stromness Museum, Orkney, photographer Rebecca Marr.

the gulf's southern coast from Victoria Harbour (where the Rosses had overwintered) to Fury and Hecla Strait (where Parry's ships had become ice-beset).[2] Rae's charts made it clear that, as Inuit living on Boothia had told the Rosses, there was no channel leading westward from that area.

In early November 1847, while Rae was still in London, newspapers reported on his findings. As he prepared to travel to Orkney to see his family, he received a letter from Dr John Richardson, who had accompanied Franklin on overland journeys in HBC territory and knew Rae's father. Richardson now hoped Rae would serve as his second-in-command on a forthcoming search for messages from or signs of Franklin and his men in the area between the mouths of the McKenzie and Coppermine Rivers.

In March 1848, following Rae's home leave, the two Scottish doctors left Liverpool for New York, from where they travelled by steamer to Montreal and reported to Sir George Simpson at his mansion. Rae made a detour to Toronto to visit Lefroy, then joined Richardson's twenty-man party at Fort Confidence, which lay at around 120°W, near the western end of the Arctic archipelago.[3] While they overwintered, Rae and Richardson agreed that, as one of their boats was already badly damaged, Rae, his Inuit guide

Paul Kane (Canadian, 1810–71), *Scene in the Northwest* – Portrait [John Henry Lefroy], *c*.1845–46, oil on canvas (overall, 55.5 x 76cm/*c*.22 x 30in), The Thomson Collection at the Art Gallery of Ontario, 2009/507; photograph © Art Gallery of Ontario.

and a group of other experienced travellers would travel to the mouth of the Coppermine River and continue the search, while Richardson and other British-based expedition members would return home.

Between June and August, Rae and his party searched in vain for messages or other indications that Franklin's ships or men had passed through the area. Rae considered crossing the strait to Woolaston Land, but as the winter weather was now setting in and Inuit reported no sightings of 'kabloonas' or their large ships, he decided to call a halt to his research and report back to Richardson and others.

In Rae's absence, Sir George Simpson had promoted him to manager of the company's Mackenzie River district, based at Fort Simpson. In late September, just as Rae was settling in, a small group arrived from HMS *Plover*, one of the two ships now searching for Franklin's ships eastwards from the Bering Strait. Rae assisted the group as much as he could, including by loaning some navigational instruments that Richardson had left behind.

As Rae had spent less time than hoped with his family on his last visit to Britain, he applied for another period of home leave. He was

Memorial to John Rae, St Magnus Cathedral, Kirkwall, Orkney; image © A. Strathie.

somewhat taken aback, however, when Simpson's response was to promote him to the role of chief factor, a role Simpson wanted Rae to combine with assisting in searches for Franklin and his men.

In April 1851, as concern for Franklin's ships and men grew, Rae led another search and survey expedition to the coastal area adjacent to the mouth of the Coppermine River. This time, as he looked over to Woolaston and Victoria Land, he realised that Dease and Simpson had not surveyed the whole area. Over the next few months, working west to east, including under the midnight sun, to avoid snow blindness, Rae and his companions charted the coast between 117°W and 101°W.

In late August, during a stop at Parker Bay (around 69°N 103°W), Rae began inspecting a piece of pine which resembled a flagstaff and lay among the debris on the shore. He noticed, to his surprise, that a rope attached to it had an Admiralty-style red 'rogue's yarn' thread running through it and metal tacks in the wood were stamped with the Admiralty's broad arrow symbol.[4] He wondered whether the wood was from *Erebus* or *Terror*, but as Inuit living on Victoria Land had already confirmed they had seen no ships, he wondered whether this and other debris might have been swept south through a previously uncharted channel which, as Franklin had envisioned, ran southwards from Barrow Strait to the south-east coast of Victoria Land.

As Rae's work for the season was done, he returned to Fort Simpson and completed reports and charts for submission to Sir George Simpson, the Admiralty (to whom he also sent some of the wood) and the Royal Geographical Society. Rae had, this year alone, covered 2,500 miles of coastline, of which 600 miles had not previously been charted.

As Sir George Simpson had finally authorised his home leave, Rae set off on the first stage of his journey to Orkney, which involved crossing Canada by river and on foot, then travelling to St Paul, Minnesota – a distance of 1,700 miles.[5] In early April 1852, John Rae arrived in London – but while he looked forward to seeing members of his family again, he was already planning another survey and Franklin search expedition.

POLAR POSTSCRIPT: In May 1852, the Royal Geographical Society awarded Rae their Founder's Gold Medal for his 'survey of Boothia under most severe privations [... and] very important contributions to the Geography of the Arctic'. Rae was by then in Orkney with his family, so could not collect his medal in person, but he confirmed he was already planning a survey of the remaining uncharted areas on and adjacent to Boothia.

16

A Graveyard on Beechey Island

Beechey Island was first charted during Edward Parry's 1819–20 expedition and lies near the junction of Barrow Strait and Wellington Channel. As no other Northwest Passage expeditions or whalers sailed as far west during the following decades, the island was little visited other than by Inuit, until those searching for Franklin's ships and men passed Prince Regent Inlet (at approximately 90°W) on Barrow Strait.

Beechey Island lies at around 75°N and was named for Sir William Beechey RA, the father of Parry's fellow officer and friend, Frederick Beechey. Its main geographical feature is a wide bay, which is sheltered by slopes that rise to around 650ft.

Sir John Barrow died in November 1848 without knowing the outcome of the Admiralty's initial three-pronged search for Franklin's ships. By summer 1850, thanks in part to incentives offered by the Admiralty and Jane Franklin, Barrow's son John (now an Admiralty official) was keeping track of a dozen search vessels: *Enterprise* and *Investigator* (under Collinson and McClure); HMS *Resolute* (under Horatio Austin); HMS *Assistance* (under Erasmus Ommanney); steamers *Pioneer*, *Intrepid*, *Rescue* and *Advance* (commissioned by American ship owner and philanthropist Henry Grinnell); *Prince Albert*, *Lady Franklin* and *Sophia* (financed by Jane Franklin and her supporters); and schooner *Felix* and motor launch *Mary*, which were commanded by 73-year-old John Ross and part-financed by Felix Booth and Hudson's Bay Company.

Ice conditions in 1850 in Barrow Strait were little better than those James Ross had experienced during his search expedition, but now, thanks to Ross, searchers had more detailed charts of the area. While expedition

Calotype image of Beechey Island graves with original grave markers of (l to r) William Braine, John Hartnell and John Torrington (in album DRO/8760/F/LIB/10/1/1); image © and courtesy of Derbyshire Record Office.

commanders waited for ice to clear from Peel Sound, parties of men in auxiliary boats landed and searched for messages from or signs of Franklin and his men. In late August, a party from Ommanney's ice-beset *Assistance* crossed from Cape Riley to Beechey Island, where they found a cairn. The cairn was well built and sturdy, but after the men checked for and found no messages they moved on to Cornwallis Island.

When *Lady Franklin*'s captain, Scottish whaling master William Penny, learned from those still on *Assistance* that only parts of Beechey Island had been searched, he mustered his own landing party. Penny, a friend of *Terror*'s assistant surgeon Alexander McDonald, had been searching unsuccessfully for messages or signs of the expedition during recent whaling seasons. He had also, for the second year running, brought with him as his ship's surgeon Robert Goodsir, a younger brother of Harry Goodsir.[1] Penny and Goodsir's 1849 search had been hampered by ice, but they hoped that this year they would make more

Beechey Island graves, showing replica grave markers, August 2023; image © A. Strathie.

progress and have news of Alexander McDonald, Harry Goodsir and their shipmates.

On 27 August 1850, Goodsir and Johan Petersen (a Danish–Inuit dog-handler and interpreter) landed on Beechey Island's debris-strewn shingle beach. Ommanney's men had checked the stone cairn, but Petersen spotted another larger cairn which, on inspection, turned out to be made from empty tin canisters marked with the Admiralty's broad arrow and the name of the supplier, Stephen Goldner.[2] Petersen also pointed out three near-identical dark objects rising from the beach.

Initially, Robert thought they looked like small huts, but as he approached he saw they were not shelters but black slabs, perhaps made from painted wood. Then he noticed three oblong mounds extending beyond the slabs. Robert's heart began thudding as he approached what he now knew must be three graves. The side of the boards facing him was blank, so he braced himself to walk round the grave-markers and read the names on them.

John Torrington, William Braine and John Hartnell had all apparently died in early 1846. No causes were given, but the lack of other graves near what appeared to be Franklin's first winter quarters suggested to Robert that his brother had survived the expedition's first winter. But as no notes or Admiralty forms had yet been found in cairns or other likely caches, searchers still had no idea whether *Erebus* and *Terror* had, after overwintering on Beechey Island, headed north up Wellington Channel or south-west towards the Canadian coast.

When search parties gathered to exchange news and compare findings, someone produced what looked like a long-handled rake, which had been found at Cape Riley. After Midshipman Clements Markham from HMS *Assistance*, Robert and others examined it, they agreed it could be a naturalist's instrument, possibly used for collecting seaweed or other samples from the seabed in shallow waters. No one was sure, although Robert knew that his brother had, shortly before leaving London, been appointed *Erebus*'s naturalist in addition to his role as assistant surgeon. Although it seemed strange that a naturalist would abandon such an instrument, it suggested that Harry Goodsir, who regularly searched for specimens on beaches in his native Fife, might have combed the Arctic shores in hopes of finding an unidentified or unrecorded species.

POLAR POSTSCRIPT: In late 1850, British newspapers reported that 'traces' of the Franklin expedition had been found on Beechey Island, but no mention was initially made of graves. Following his two searches with Captain Penny, Robert Goodsir completed his medical studies in Scotland, then immigrated to Australia. His account of his visit to Beechey Island was published in a newspaper in 1880. Two years later, he returned to Edinburgh, where he lived with his sister Jane, and where, following his death in 1895, he was buried in the city's Dean Cemetery. William Penny returned to Barrow Strait in 1851, but after Captain Austin declined to loan him an Admiralty steamship for a search of Wellington Channel, he resumed his whaling activities.

17

Eleanor Gell's 'Franklin Search' Collection

During the late 1840s, Eleanor Franklin's stepmother, Lady Jane Franklin, became nationally and internationally famous for her advocacy and financial support of Franklin search expeditions. By contrast, the name of Franklin's only child, Eleanor Franklin, was rarely mentioned in newspapers. The material she and her husband, Reverend Philip Gell, collected during the period of the Franklin search expeditions, along with related correspondence, provide a more personal view of what was sometimes a very public story.

The Gells' material relating to the Franklin search expeditions was mounted in scrapbooks (ref. D8760/F/LIB/10/1/1-2) which, along with other records relating to Eleanor and her Franklin relatives, are part of the Gell Family Collection at Derbyshire Record Office, Matlock.

When Eleanor Franklin married Philip Gell on 7 June 1849, her father was not present. He had, however, in his letter of Sunday, 6 July 1845, written in Disko Bay, Greenland, anticipated her 'united happiness' with Gell, whom she had first met in Hobart after her father appointed him as principal elect of Van Diemen's Land's first higher education institute, Christ's College.[1]

Shortly after the Gells married, James Ross and John Richardson returned from the Arctic, sorely disappointed to have found no indication as to where her father's ships might be. Eleanor had, in 1848, sent letters to him and Crozier via Ross and Richardson, but they remained undelivered, as did one she sent her father the following year.[2]

During 1850, as further search expeditions departed and returned, the Gells' collection of newspaper reports mounted. That year Eleanor gave

Letter from John Franklin to his daughter Eleanor Gell, 6 July 1845, DRO/D8760/F/EG//1/1/15 (full text in letter 149, Potter et al., *May We Be Spared*, see Bibliography); image © and courtesy of Derbyshire Record Office.

birth to a daughter. She no longer wrote to her father, but a calotype print she was given of the three graves found on Beechey Island was a sober reminder of the hardships her father and his companions faced. Eleanor's stepmother Jane and cousin Sophy (now effectively the latter's full-time companion) continued to write to her father regularly, but by mid-1852, Eleanor had accepted that he might well be dead.[3] As her husband was on a modest clergyman's stipend, Eleanor was in no position to sponsor search expeditions, particularly as the funds her mother had left her remained under her father's (and now her stepmother's) control.[4]

As the Gells struggled to raise three children on his stipend and a small allowance from Jane Franklin, Gell applied for post of bishop of a new diocese in Christchurch, New Zealand. But before the outcome of his application was known, the Admiralty announced that unless they received proof by 31 March 1854 that expedition members were alive, they would be presumed dead, and their names would be removed from Admiralty payrolls.[5]

In October 1854, *The Times* published extracts from a report that John Rae had submitted to HBC officials based on his latest Franklin search. Rae had, it seemed, met members of the Inuit community, one of whom described a large group of 'kabloonas' who had apparently died some years ago from starvation. Rae's enquiries suggested that knowledge of the story was widespread locally and when the Inuit showed him monogrammed cutlery with initials that corresponded to expedition members' names, he became increasingly convinced the Inuit had seen Franklin's men.

Rae was also told that bodies had been found on an island near the estuary of the Great Fish River – which, given Rae had recently charted a narrow strait separating Boothia from King William Land, suggested that they were referring to King William Island. Rae suggested in his report to HBC officials that it was disturbing that, based on what the Inuit had seen, the starving men may have been driven to what he described as 'the last resource – cannibalism'.

Rae had not expected either HBC or the Admiralty to release his report to *The Times*, where it was published unedited. He became the object of Jane Franklin's anger on the grounds that his use of the word 'cannibalism' was an insult to her deeply religious husband. After she enlisted the support of Charles Dickens, Rae was also chastised for relying on the word of Inuit people and for not visiting the sites of the remains in person. But after the *Illustrated London News* published drawings of artefacts brought back by Rae, and Dickens acknowledged that John Rae had prepared a

Press cuttings showing the 1829 medallion portrait of John Franklin and daguerreotype of John Rae (studio of Richard Beard), from album DRO/D8760/F/LIB/10/1/2; images © and courtesy of Derbyshire Record Office.

less clinical version of his report for public consumption, the furore began to die down.[6]

During 1855, after John Franklin's will was proved, the Gells and Jane Franklin reached a financial settlement and, thanks to Eleanor's network of Franklin, Cracroft, Lefroy and other cousins, family harmony was restored.[7] Jane was, however, still determined to establish what had happened to her husband, so she raised funds to purchase the steam yacht *Fox* so that Leopold McClintock could return to the Arctic for a final search.

In September 1859, McClintock returned to England with more relics and messages found in a cairn on the west coast of King William Island. The messages were all written on an Admiralty pro forma but bore different dates. The first, dated 28 May 1847, suggested that although *Erebus* and *Terror* had been ice-beset since 12 September 1846, all was still well. The second message, dated 25 April 1848, stated that Franklin had died on 11 June 1847, and that by the date of the message another twenty-four officers and men had died – the only one named was Graham Gore, who had deposited an earlier version of the note in another cairn. The later message was signed by James Fitzjames, but a brief addendum signed by Crozier suggested that the remaining 105 men were heading south towards the estuary of the Great Fish River.

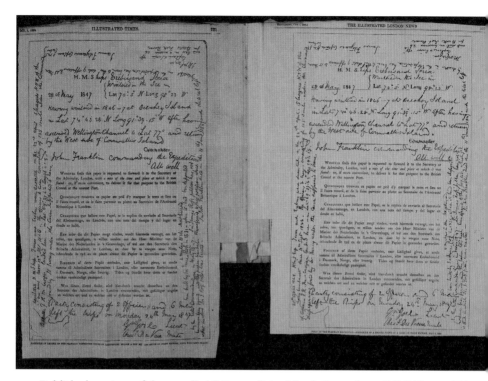

Published versions of the so-called 'Victory Point Note', from album DRO/D8760/F/LIB/10/1/2; image © and courtesy of Derbyshire Record Office.

McClintock and Rae were duly rewarded for their efforts. Rae received an Admiralty award of £10,000, while McClintock was knighted, shared £5,000 with his men and was presented with the Royal Geographical Society's 1860 Patron's Medal. Jane Franklin was awarded that year's Founder's Medal in recognition of her husband's achievements and her own 'self-sacrificing perseverance in sending out expeditions'.

The Gells had expected to be living in New Zealand by 1860, but Philip Gell's appointment to the bishopric of Christchurch failed to materialise.[8] As Eleanor had never visited Orkney, her father's last British port of call, the Gells wrote to Rae. He would, he apologised, be away from home, working on a survey of the North Atlantic with Leopold McClintock, but enclosed letters of introduction to three of his Orcadian friends.[9] In the event, the Gells, whose seventh child was barely a year old, spent summer 1860 in Tredunnock, South Wales, where Philip Gell was covering for the local vicar. Eleanor, whose constitution (like that of her mother) was not strong, contracted scarlet fever while there and died on 30 August.

Eleanor Gell's death was widely reported, but most announcements were short and, perhaps inevitably, referred to her in terms of her father. But she was much loved and much missed by members of her immediate and wider family and by her husband's parishioners and others who had benefited from her generosity and good works.

POLAR POSTSCRIPT: Eleanor Gell was buried in the graveyard of Tredunnock's parish church. She never knew where her father was buried, but the year after her death a statue of her father was unveiled in the marketplace at Spilsby, Lincolnshire, near the shop where he had been raised and where many of Eleanor's relatives still lived. Philip Gell never remarried, but he and his children kept papers and letters relating to the search for his beloved wife's father, including the precious scrapbooks of memorabilia which passed down the generations.

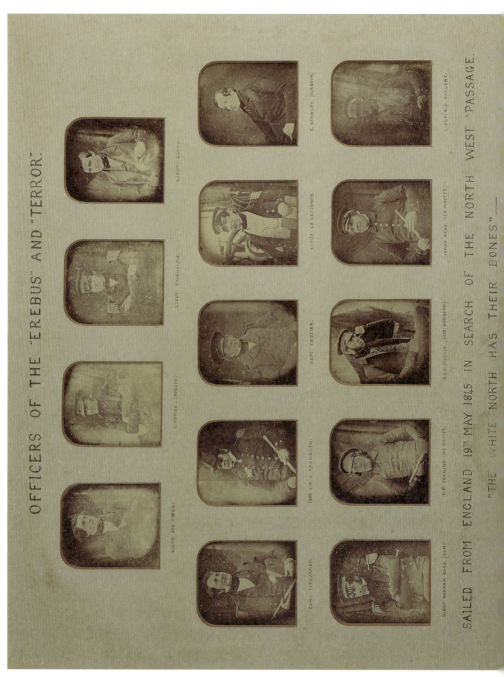

As per DRO catalogue: Top: Lieut. Des Voeux, C. Osmer (Purser), Lieut. Fairholme, Lieut. Couch;
Middle: Com. Fitzjames, Capt. Sir J Franklin, Capt. Crozier, Lieut. Le Vesconte, S. Stanley (Surgeon)
Bottom: Lieut. Graham Gore (Com.), H.F. Collins (Ice Master), H.D.S. Goodsir (Asst. Surgeon),
James Read [Reid] (Ice Master), Lieut. R.O. Sargent
Mounted set of photographic prints of Beard's daguerreotypes of HMS *Erebus* officers and
others and Francis Crozier (HMS *Terror*), c.1851, D8760/F/LIB/8/1/5;
image © and courtesy of Derbyshire Record Office.

Part IV

A New Start

In 1866, a statue of John Franklin was erected, at government expense, in Waterloo Place, off Pall Mall. At Jane Franklin's instigation an inscription on the statue's plinth suggested that Franklin and his men had completed 'the discovery of the North-West Passage'. Although there were other contenders for the title of 'discoverer of the Northwest Passage', it was generally agreed that an expedition which had resulted in the deaths of all its members and thirty of those who searched for them was a national tragedy.

By 1870, however, most naval officers who had participated in Arctic expeditions over the past five decades or so had died or retired. The Admiralty, which had lost several ships and spent large amounts of money during that period, ceded its leadership role in voyages of geographical and scientific exploration to the Royal Society and Royal Geographical Society but agreed that, subject to operational considerations, Admiralty officials would consider requests to loan vessels or to second officers.

In 1872, Royal Society officials and scientists from the University of Edinburgh were planning a major oceanographic survey in far southern latitudes on HMS *Challenger*, the former flagship of the navy's Australia Station. Meanwhile, the Royal Geographical Society (President: George Back; Hon. Secretary: Clements Markham) established an Arctic Committee. While a more or less viable route through the Northwest Passage had now been charted, John Rae and other members of the committee were invited to consider the best means of reaching the North Pole.

In 1874, Royal Academician John Millais exhibited a new painting entitled 'The North-west Passage: It can be done, and England ought to do it' at the Academy's summer show. As the Northwest Passage was now charted, some were puzzled by the title. Before long, however, 'The North-west Passage' became one of Millais's most popular and regularly reproduced works and contributed to a revival in the public's interest in polar exploration.[1]

18

A Photograph of Antarctic Icebergs

In early February 1874, as HMS *Challenger* approached the Antarctic Circle, photographer Frederick Hodgeson and his shipmates saw their first free-floating icebergs. Over a two-week period, when conditions allowed, Hodgeson set up his tripod and exposed glass plates of the icy monoliths. Several of the photographs, believed to be the first of Antarctic icebergs, were used to illustrate expedition reports and lectures.

Glass plates with Hodgeson's images of icebergs were indexed as Nos 279–87 following the expedition. The plates and prints of them are in the collection of the Natural History Museum, London. Hodgeson joined *Challenger* in South Africa in December 1873, but little is known of his work before or following the expedition.[1]

HMS *Challenger*'s commander, Captain George Nares, had experience both of oceanic surveying and (from serving on HMS *Resolute*) polar conditions. As the expedition was largely a scientific enterprise, Nares selected officers with similar experience, so they could assist chief scientist C. Wyville Thomson, oceanographer John Murray and others as they collected data on the world's deepest oceans and basins and dredged samples from ocean beds. *Challenger* had, like James Cook's and James Ross's vessels, been fitted out with scientific laboratories including, for the first time on an Admiralty vessel, a fully equipped photographic laboratory, where glass plates could be developed and the results printed immediately.

William Abney, head of the Royal Engineers' School of Photography, assisted with the designing and fitting out of the laboratory and arranged for the secondment of one of his best students, Corporal Caleb Newbold, to the expedition. Thomson and his fellow scientists were impressed by

Icebergs (photograph no. 280 from the Challenger Collection, ID 12565); image © and courtesy of Trustees of the Natural History Museum, London.

Newbold's work, which included photographing specimens and scientific drawings. But while *Challenger* was being overhauled in Simonstown, near Cape Town, Newbold disappeared.[2]

Thomson knew that the combination of the ship's motion, damp sea air and extreme temperatures, which affected photographic equipment and chemicals, made this a difficult assignment, but he soon secured the services of Frederick Hodgeson, a local commercial photographer.[3] Hodgeson initially suffered from seasickness, but as *Challenger* approached the Antarctic Circle, he exposed several glass plates as tabular and sea-worn icebergs passed the ship. As Hodgeson's images would be monochrome, however, the expedition secretary-cum-artist John Wild, sub-lieutenant Herbert Swire, able-seaman John Arthur and others were on hand to sketch icebergs and to attempt to hold in their mind's eye the different shades of blue and aquamarine streaks running through the ice.[4]

After completing the southernmost leg of the oceanographic research programme, Nares set course for Australia, where *Challenger* remained for almost three months. Following brief stops in New Zealand, Fiji and other Pacific islands, *Challenger* reached Hong Kong where, after two

'Ice Floes on Choppy Water', watercolour (1874), Herbert Swire (ref. H2017.82/1); image © and courtesy of State Library Victoria.

'HMS *Challenger* in Collision with Iceberg, Feb'y 24' [1874], John Arthur (ref. H31463); image © and courtesy of State Library Victoria.

years away, everyone enjoyed celebrating Christmas and welcoming in 1875. While *Challenger* was in port, however, Nares was notified that he had been appointed commander of the Admiralty's first Arctic voyage of exploration since 1845.

Before Nares and Lieutenant Aldrich (who was also to join the Arctic expedition) left Hong Kong, Captain Frank Thomson of the China Station took command of *Challenger*. For unknown reasons, Frederick Hodgeson also left the ship at Hong Kong, but the China Station's commander suggested that Jesse Lay, a clerk on a naval hospital ship who was a keen amateur photographer, would be a suitable replacement.

Challenger returned to Britain in May 1876, by which time Thomson, Murray and their fellow scientists had, in approximate terms, sailed 68,000 nautical miles, recorded 500 depth soundings, set up 400 sampling stations and recorded 5,000 previously unknown species, including marine creatures living at depths well in excess of any previously recorded. In addition, rock and soil samples from dredging near the Antarctic Circle suggested that a sizeable land mass lay not far south of their course.

A two-volume *Narrative of the Cruise*, the first of fifty volumes of reports generated from the expedition's findings, was published in 1885 and advertised as being 'profusely Illustrated with Chromo-lithographic and Photographic Plates, Maps, Diagrams, Woodcuts'. Among the lithographic and photographic plates were several of Frederick Hodgeson's photographs of icebergs and Caleb Newbold's photographs of sub-Antarctic penguin colonies. Several plates were credited to expedition artist John Wild, whose name also appeared as a member of staff, but photographers Caleb Newbold, Frederick Hodgeson and Jesse Lay received no official recognition for their efforts in recording the work of one of Britain's longest and most successful scientific expeditions.[5]

POLAR POSTSCRIPT: The Arctic Medal remained, despite James Ross's previous protestations, the only decoration dedicated to those who explored polar regions. Following the *Challenger* expedition, scientist John Murray arranged, at his own expense, for a *Challenger* Medal to be designed, cast and presented to expedition members and those who assisted with scientific reports. In recognition of the expedition's largely oceanographic mission, the reverse of the 1895 medal showed Neptune holding a trident (*Challenger*'s crest) and a trawling net.[6]

19

A Menu for a Banquet

This menu was produced for a banquet held in Southsea's Assembly Rooms on 30 November 1876, in honour of the safe return of Captain George Nares and his companions from their Arctic expedition on *Alert* and *Discovery*.[1] As Nares's ships had sailed from Portsmouth and passed along Southsea's seafront, he regarded this as a homecoming, particularly as his wife's family were from the area. An attempt to reach the North Pole from Ellesmere Land had failed, but a sledging party led by Albert Nares reached a Farthest North of 83° 20'N and previously uncharted areas had been surveyed.

The menu for the seven-course banquet when open measures 9¼in by 3½in; it bears the city crest of Portsmouth and, on the back, small photographic prints of *Alert* and *Discovery* (private collection).

When George Nares returned from his Arctic expedition in October 1876, there was talk of him being knighted for his services, but Admiralty officials raised concerns regarding outbreaks of scurvy, particularly during sledge journeys. Following his return, Nares kept his counsel on the matter, but on the evening of 30 November, when he was guest of honour at a lavish homecoming banquet hosted by Portsmouth's Mayor William Pink, he lowered his guard.

Southsea's seafront Assembly Rooms were decorated for the occasion with flags, banners with 'icicled' lettering and shields bearing the names of famous Arctic explorers. Nares's wife and members of her family were present, as was Leopold McClintock, who was now Superintendent of Portsmouth Dockyard – in which capacity he had helped Nares prepare for his Arctic voyage.[2]

Following the seven-course banquet, guests raised their glasses to Nares and his fellow explorers. In his response, he explained that, as he was now

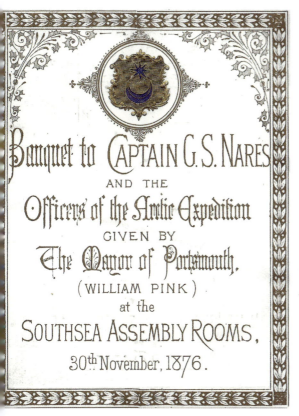
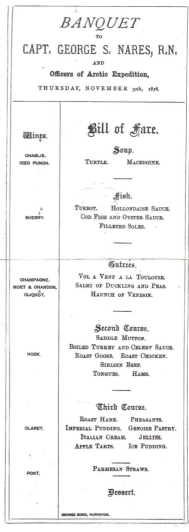

Menu card front, back (with photograph of ships) and inside (banquet menu); images © A. Strathie, author's collection.

HMS *Alert* and *Discovery* passing Southport, *Illustrated London News*, 5 June 1876; author's collection.

among friends, he wanted to correct misapprehensions resulting from some comments attributed to him. He had never, he explained, suggested that reaching the North Pole was 'impracticable'. The Farthest North set by Commander Albert Markham and his sledge party had, he explained, been achieved in brutally low temperatures, which had rendered the men's outdoor clothing inadequate and contributed to cases of scurvy and frostbite and the unfortunate deaths of several men.[3] After Markham's sledge party had returned to the ships, however, it had become clear that attempting to reach the North Pole in spring 1877 would indeed be 'impracticable'. As to reported issues with sledges, the total weight hauled by Markham and others was not (as had been suggested) 400–500lb, but around 240lb, similar to weights hauled by the men during McClintock's Franklin search expedition.

During the evening, Mayor Pink and his guests cheered their local hero and his companions numerous times. Notwithstanding the apparent controversies, Nares was duly knighted and honoured by the Royal Society and Royal Geographical Society. Nares had decided he would leave reaching the North Pole to others but noted in the introduction to his expedition report that 'no Arctic expedition can ever depart without a full equipment of sledges, any more than an ordinary ship can sail on a voyage without her proper complement of boats [given that] sledges are the only means of locomotion in these [polar] regions'.

Albert Markham, in his personal expedition narrative, *The Great Frozen Sea*, described his sledge journeys in more detail and described the officers' efforts to maintain their men's morale, including by naming sledge dogs and sledges for heroes, relatives or friends. Markham named one for his cousin, Clements, who had, in his capacity as Royal Geographical Society Secretary, accompanied the expedition as far as Baffin Bay. Markham also suggested that sledges, whether pulled Inuit-style by dog teams or by men, were a reliable means of travel in polar regions. He also felt that sledge flags, popularised by Horatio Austin and Leopold McClintock during their Franklin search expeditions, boosted morale and 'added to the spirit and gaiety of the scene'.[4]

While Nares agreed with Markham that morale was important, he suggested during the Southsea banquet that it remained the responsibility of the navy's 'medical faculty' to consider the 'terrible plague' of scurvy, which had put paid to hopes of reaching the North Pole during what had been a well-organised and well-equipped expedition. If scurvy could be eradicated, that would certainly be a cause for celebration.

POLAR POSTSCRIPT: In 1878, following a break from active service, Nares resumed command of HMS *Alert*, this time for a survey of the Strait of Magellan. He never returned to polar regions, but he was regularly consulted by Admiralty and Royal Geographical Society officials and, in the early twentieth century, by a new generation of explorers.

20

Cornelius Hulott's *Resolute* Box

This hand-carved oak box is made from timbers from HMS *Resolute* and belonged to Cornelius Hulott, who served as captain's coxswain on HMS *Investigator* during Robert McClure's 1850–54 Franklin search expedition. Hulott was born in 1830 on the Isle of Sheppey in Kent. He was the youngest of seven children of ex-French naval officer Maurice Hulott, who was born in Martinique and was of African descent. After Maurice Hulott's ship was captured during the Napoleonic Wars, he was imprisoned and discharged into the British Navy, which entitled his children to apply for places at Greenwich Hospital School. Cornelius Hulott joined the navy aged 12 and travelled widely before joining *Investigator*. The little *Resolute* box was treasured by Hulott and served as a memento of what had been one of his most exciting commissions, but one which had changed his life in many ways.

The box (around 3in diameter) was kept by Hulott until his death. He appears to have used it to store a special 1797 'cartwheel' penny. Hulott and his wife had no children, but the box is now owned by Hulott's great-great-grandniece, Lynne Flatman.

In July 1879, while Cornelius Hulott was working as a rigger in Chatham Dockyard, HMS *Resolute* was towed into No. 7 dock, where she was broken up. Hulott had never served on *Resolute* but, as he told workmates, he had spent nearly a year on her when his own ship, HMS *Investigator*, was abandoned in 1853 after becoming trapped in ice while searching for John Franklin's ships. Hulott expected to sail home on *Resolute* but, when she became frozen in too, he returned to Britain on a supply ship.

Cornelius Hulott's *Resolute* box; image © and courtesy of Lynne Flatman.

He thought *Resolute* and *Investigator* were both lost for good, but *Resolute* eventually drifted to Baffin Island, from where American whalers sailed her to New London, Connecticut. In 1856, after *Resolute* was refurbished, the American government presented her to Queen Victoria. Lady Franklin and her American supporter, Henry Grinnell, thought *Resolute* should return to the Arctic and start looking for Franklin again, but *Resolute* stayed in British waters until she was brought to Chatham for breaking up.

As *Resolute* was broken up, Chatham's head carpenter, William Evenden, and his team began using some of the timbers to make a handsome desk. The desk was carved with portraits of the queen and the American president and scenes showing *Resolute* trapped in ice, and would be presented to President Rutherford Hayes as a gift from the queen.[1] The carpenters

'Sledging over Hummocky Ice', Samuel Cresswell, from *A series of eight sketches in colour ... of the voyage of H.M.S.* Investigator *(Captain McClure)* ...; image courtesy of New York Public Library (Digital Collections).

made a smaller desk for Henry Grinnell and his wife and a writing table for the queen, but they also found time to make a little *Resolute* box for Cornelius, who told them tales about his time in the Arctic.

Hulott had been promoted to captain's coxswain after joining *Investigator*. His captain, Commander Robert McClure, who had been first lieutenant on *Investigator* on James Ross's Franklin search expedition, was known for being headstrong. After losing contact with his sister ship *Enterprise*, he decided (against orders) to enter the Bering Strait and head east alone. As *Investigator* passed the mouth of the Mackenzie River, McClure gave orders to enter a northbound channel he thought might lead to Melville Island on Barrow Strait – the furthest point Edward Parry had reached on his first attempt to traverse the Northwest Passage. When *Investigator* became icebound in a narrow channel on the way north, McClure decided that, rather than turn back, he would overwinter and hope the ice would clear early in spring 1851.

During a bitterly cold winter, Hulott and his shipmates formed sledge parties and went hunting, but while that meant they had fresh meat, their sledging rations were not sufficient.[2] Dr Armstrong and the assistant

'Critical Position of HMS *Investigator* …', Samuel Creswell; image courtesy of New York Public Library (Digital Collections).

surgeon doled out anti-scorbutic citrus juice and showed men how to revive frostbitten feet by gently rubbing them between warm hands. The doctors also checked men for signs of scurvy and, when necessary, extracted painful teeth.[3]

In late May 1851, during a mild spell, Hulott joined a sledge party for a long trek to an Inuit settlement, where McClure and translator-cum-pastor Johan Miertsching hoped to establish whether there had been any sightings of Franklin's ships or men.[4] As they crossed some slushy ice, Hulott's tight-fitting canvas boots became sodden, but he had no other footgear, so he kept them on overnight.

By the next morning, his canvas boots had shrunk and frozen round his feet, but as he marched, the boots thawed again – but when he next took his boots off, he realised that most of his toes were still frozen solid. By now, Hulott's feet were so painful that his companions suggested he walk beside the sledge.

After McClure inspected Hulott's feet, he rubbed concentrated spirit on them – which made them more inflamed than ever. Soon, Hulott's feet had become gangrenous. He rested at the Inuit settlement, but when

Desk made from timbers of *Resolute* and presented by Queen Victoria to the US President (from Frank Leslie's *Illustrated Newspaper*, 11 December 1880); image courtesy of Library of Congress.

it was time to return to the ship, his companions decided that as he was slight of build they would strap him to a sledge and pull him all the way back. When Dr Armstrong saw Hulott's feet, he was so shocked that he sent him straight to the sick bay. The kindly doctor tried to save his patient's badly damaged toes, but despite his best efforts, he had to amputate all of the toes on Hulott's right foot and his big toe and part of the second toe on his left foot.

In April 1852, McClure led a sledge party to Parry's Winter Harbour, where he found a note suggesting that other ships were not far away and left his own note. There was no sign of a thaw that summer or in spring 1853. The sick bay was full and those who were not yet sick were half-starved. So, when Lieutenant Bedford Pim of HMS *Resolute*, guided by McClure's note, arrived and told them that *Resolute* and other Franklin search vessels were at Parry's winter quarters and preparing to leave for England, there was much jubilation and relief.

By August, Hulott and his companions were aboard *Resolute* and heading east, but before long, the ice froze again, leaving *Resolute* trapped.

She was still there in April 1854, so Hulott and his *Investigator* shipmates abandoned ship for a second time and boarded *North Star* and other vessels, which brought them back to England.

When Hulott reported back for duty, doctors declared him unfit for regular service due to his damaged feet. McClure was court-martialled for abandoning *Investigator*, but was pardoned, promoted and knighted and awarded half of a £10,000 reward for traversing the Northwest Passage.[5] In due course, Hulott received £52 2s 10d as his share of the £10,000 reward, an Arctic Medal and the offer of a permanent job as a rigger at Chatham Docks. While Hulott's time in the Arctic had been difficult, he did not feel particularly hard done by, so he cherished his little *Resolute* box and named his house 'Arctic Villa'.

POLAR POSTSCRIPT: When Cornelius Hulott died in 1899 at the age of 69, the *Morning Post* published an article headed 'Death of an Arctic veteran'. The journalist referred to Hulott's father being from 'the West Indies' and to Hulott's own 'somewhat adventurous career' and 'hairsbreadth escapes' in the Arctic.[6] The *Resolute* desk remains in the White House, where it is regularly used by serving US presidents.

21

'On Board *Eira*': From the '*Eira* 1880' Album

This photograph was taken on Monday, 12 July 1880, during the maiden voyage of Benjamin Leigh Smith's custom-built research and exploration vessel *Eira*. Leigh Smith, who financed all his own expeditions, was well respected by Royal Geographical Society officials and within the exploration community. Although he was known for his dislike of publicity, he agreed to serve as a pall-bearer at Jane Franklin's funeral in 1875, along with Leopold McClintock, John Barrow and three admirals.

The photograph was taken by William 'Johnny' Grant, expedition photographer on George Nares's Arctic expedition. It shows (left to right) Captain David Gray (*Hope*), Benjamin Leigh Smith (*Eira*), Dr Arthur Doyle (*Hope*), Captain William Lofley (ice master, *Eira*), Captain John Gray (*Eclipse*), Dr Robert Walker (*Eclipse*) and Dr William Neale (*Eira*). This and other photos from the same season are in an '*Eira* 1880' album (twenty photographs and a map) held by Hull Museum (ref. 2005.5367).

In May 1880, Benjamin Leigh Smith's new 125ft steam yacht *Eira* was launched from Peterhead, Aberdeenshire. Although Leigh Smith had previously chartered vessels from Hull or Dundee for expeditions to Svalbard and Jan Mayen Land, he had, with advice from his old friend, whaling master David Gray, commissioned a new ship which he hoped would be capable of reaching the North Pole.[1]

On Sunday, 11 July, between Jan Mayen Island and Spitsbergen, Leigh Smith spotted *Hope* and *Eclipse*, the ships of David Gray and his brother John, and as he had letters for David Gray, the three ships anchored to the same ice floe. Leigh Smith was disappointed when David Gray suggested there was understandably little chance of reaching the North Pole this season.

'On board *Eira*' (see text for full details), from '*Eira* 1880' album; image © and courtesy of Hull Maritime Museum/Hull Culture & Leisure Ltd.

The following evening, Leigh Smith invited the Grays and their ships' doctors for farewell after-dinner champagne and cigars on *Eira*. It was still light on deck, so Johnny Grant set up his camera and photographed the three captains, three doctors and *Eira*'s ice master in near daylight. Leigh Smith was heading north the following day, and he and Neale wished *Hope*'s doctor, 21-year-old Arthur Doyle, all the best with his continuing medical studies at Edinburgh University.

Leigh Smith followed David Gray's advice and bypassed Spitsbergen, and by mid-August he had reached Franz Josef Land, where he found a sheltered harbour on an uncharted island which he named Northbrook Island. Over the next few weeks, Leigh Smith and his team surveyed and charted other geographical features, including Nightingale Sound (named for his cousin Florence) and, at the far end of Northbrook Island, Cape Flora, a peninsula where Leigh Smith could botanise to his heart's content.[2]

Eira Harbour and Bell Island (same album); image © and courtesy of Hull Maritime Museum/Hull Culture & Leisure Ltd.

When Leigh Smith submitted his season's results to his friend Clements Markham at the Royal Geographical Society, Markham was so impressed with the results that he offered to edit Leigh Smith's and Neale's records into a formal report. In January 1881, to his great surprise, Leigh Smith learned that he had been awarded the Royal Geographical Society's Patron's Medal, which he asked Markham to collect on his behalf.[3]

In June 1881, Leigh Smith returned to Franz Josef Land, but this time he encountered pack ice at 72°N and found his previous mooring on Northbrook Island blocked. Rather than risk becoming frozen in, he headed south to Bell Island and assembled 'Eira Lodge', a storehouse and shelter which he thought might come in useful later.

On 16 August, after mooring to offshore fast ice at Cape Flora, Leigh Smith and Neale went ashore and botanised on the cliffs. Ice master Lofley had noticed more pack ice forming, but he and Leigh Smith were both shocked when, on 21 August, in calm, sunny weather, the tide swept huge quantities of sea ice onto Eira's seaward flank. As the incoming ice forced

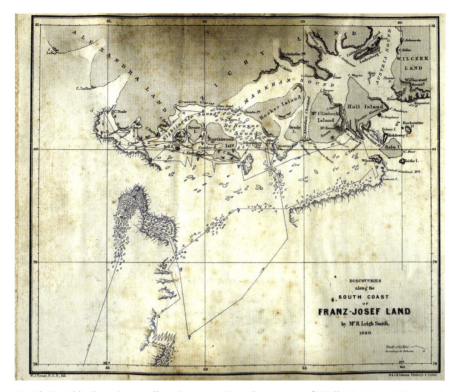

Leigh Smith's chart (same album); image © and courtesy of Hull Maritime Museum/ Hull Culture & Leisure Ltd.

Eira against the fast ice, her hull was pierced. As water gushed in, a team of men began pumping it out, while others lowered stores, tents and other equipment, fuel and auxiliary boats onto the fast ice.

Before long, *Eira* was lying 11 fathoms deep on the seabed, with only the tips of her masts to show where she had been. With Bell Island and Eira Lodge too far away, the men gathered driftwood and other materials for a makeshift shelter and moved stores and equipment higher up the beach. As they settled into 'Flora Cottage', a 38ft by 12ft lodge, the men went hunting, but as animals and birds headed south and fresh meat stocks dwindled, Leigh Smith and Neale kept an eye on rations – with the exception of the Christmas and New Year period. To keep his men occupied, Leigh Smith insisted on shipboard discipline and routines but, thanks to a salvaged musical box and occasional theatrical entertainments, the evenings did not hang too heavily.

During early 1882, glorious displays of aurora alternated with freezing snowstorms and regular gales, but as temperatures rose, the wildlife

'Foundering of the Eira' (from *Illustrated London News*, 9 September 1882); image courtesy of Freshwater and Marine Image Bank, University of Washington.

returned. Fresh meat had obvious health benefits, but it also allowed them to conserve non-perishable provisions for their forthcoming boat journey to Novaya Zemlya, where Leigh Smith knew whalers gathered each season.

On 21 June, just over a year after leaving Peterhead, Leigh Smith and his men tucked into a farewell tea at Flora Cottage, then boarded four reinforced and heavily laden auxiliary boats. For the next six weeks, usually in atrocious weather, they battled through pack ice, dragged boats over ice floes and, in occasional open water, rowed or sailed under the midnight sun until they reached the pack edge. When they camped, they were sometimes visited by bears and to obtain fresh meat they had to catch and kill walruses. But in late August, they made a final effort and reached a sheltered bay on the south-east coast of Novaya Zemlya, where they collapsed with exhaustion.

Before long, Leigh Smith and his men were aboard John Gray's *Hope*, which had been chartered for an '*Eira* Search and Relief Expedition', which had been funded by the Admiralty, the Royal Geographical Society and concerned individuals.[4] *Hope*'s temporary captain, Allen Young, had also searched for John Franklin and George Nares in the Arctic and was delighted to find Leigh Smith's men scurvy-free and, all things considered, in relatively good spirits.

When *Hope* reached Aberdeen, Leigh Smith and his men were cheered ashore. Leigh Smith was clearly exhausted by his ordeal, but while still recovering, wrote a letter of thanks to members of the *Eira* Relief Committee, thanking them for their 'noble efforts' and apologising for the fact that, due to the loss of *Eira*, the season's scientific results were 'almost nil'.[5]

Leigh Smith was, it seemed, being too modest. After Clements Markham compiled records salvaged by Leigh Smith and *Eira*'s doctor-cum-scientist William Neale into a report, it was sufficiently comprehensive and interesting for Neale to present at a Royal Geographical Society meeting. While Leigh Smith hated publicity, he seemed grateful for the efforts that had been made to save him and his companions and perhaps felt that, on this occasion at least, having his picture and that of his beloved *Eira* in magazines was a small price to pay for the safe homecoming of him and his shipmates.

POLAR POSTSCRIPT: In 1883, Dr Arthur Doyle, *Hope*'s erstwhile ship's doctor, published a short story entitled *The Captain of the 'Pole Star'*, which was set on a whaler and narrated by a young ship's doctor. In 1893, by which time he was known as A. Conan Doyle, the creator of the fictional detective Sherlock Holmes and his associate Dr Watson, he published an article entitled 'The Glamour of the Arctic'. Those to whom Leigh Smith gave copies of the 'On Board *Eira*' photograph, might have noticed that *Eira*'s Dr William Neale sported a deerstalker and hooked pipe very similar to those favoured by Sherlock Holmes. Dr Neale might also have noticed that Holmes's friend, Dr Watson, had, like him, studied medicine at the University of London in the late 1870s.

22

Illustrated London News Front Page

On 17 June 1896, British explorer Frederick Jackson saw a tall figure heading across the ice towards his expedition base on Cape Flora, Franz Josef Land. As Jackson went to meet the clearly exhausted traveller, he recognised Norwegian explorer Fridtjof Nansen, who was, as far as Jackson knew, attempting to reach the North Pole in his expedition ship *Fram*. Almost three months later, *Illustrated London News*, the world's first illustrated magazine, dedicated its front page to an image of Jackson and Nansen shaking hands. The image was captioned 'I am awfully glad to see you', and recalled the famous occasion, twenty-five years previously, when Henry Stanley and explorer David Livingstone met in Africa.

Illustrated London News (founded 1842) was a pioneer both in illustrations and in the printing of photographs. Daily newspapers, whose larger circulation required faster printing presses, did not regularly print photographs until many years later.

Frederick Jackson and Fridtjof Nansen first met during the latter's British tour, which followed his 1888–89 crossing of Greenland. During a subsequent Royal Geographical Society lecture, Nansen outlined his plans for reaching the North Pole on *Fram*, his new custom-built expedition ship. Following the lecture, Jackson wrote to Nansen asking if he could join his next expedition. Nansen agreed to meet but explained that, as he was campaigning for Norwegian independence, all expedition members would be Norwegian.[1]

Jackson began planning his own expedition to the North Pole by way of Franz Josef Land, where he would survey areas that Leigh Smith and others had not charted.[2] Royal Geographical Society Secretary Clements

Front page of *Illustrated London News*, 12 September 1896; author's collection.

Markham was enthusiastic, as was publisher Alfred Harmsworth, who agreed to fund the expedition, subject to it being called the 'Jackson-Harmsworth Expedition' and Jackson giving Harmsworth publishing rights to Jackson's reports and photographs.

When Jackson left England in July 1894, Nansen had been away for a year, but as Nansen would let *Fram* freeze into the ice before 'drifting' to the Pole, he was expected to be away for several years. As Jackson did not want his steam-assisted Scottish whaler *Windward* to be frozen in, ice master John Crowther would return south each winter, leaving Jackson, merchant mariner Albert Armitage, scientists Dr Reginald Koettlitz and David Wilton, and others to overwinter in Franz Josef Land.

After a stop in Russia to collect Siberian ponies, dogs and furs, they reached Franz Josef Land in early September. Pack ice made unloading slow and by the time they unloaded everything and erected their winter quarters at Cape Flora, the weather was so poor that *Windward* could not leave.

During the spring of 1895, Jackson, Armitage and Karl Blomkvist explored uncharted territory, accompanied by two ponies pulling four laden sledges. By midsummer, soft surfaces slowed travel, but the ice remained intractable, and it was only in July that, after the men broke up ice with axes and saws, *Windward* floated free and returned south, carrying Jackson's reports to Harmsworth and Clements Markham and everyone's home mail. During high summer, Jackson and his companions explored the area in kayaks and a 25ft whaleboat, before resuming hunting trips to ensure they had sufficient fresh meat in store for the winter.

In March 1896, Jackson, Armitage and Blomkvist headed north with a pony and dog sledges. As Jackson looked ahead, he saw sea ice where some charts showed land and realised that Franz Josef Land was probably an archipelago rather than, as he and Leigh Smith hoped, a gateway to an overland route to the North Pole.

Over the summer, Jackson and his companions kept busy, but headed west, adding more islands and capes to their charts and photographed the area. By now, however, everyone was longing for *Windward* to bring their first news of the outside world for two years.

On 17 June, Armitage came running over to tell Jackson that he had seen, through his field glasses, a man walking across an ice floe near Cape Flora. Jackson, after checking this for himself, fired a pistol in the air to attract the attention of the man, who looked as if he was wearing skis. Fridtjof Nansen was clearly glad to see Jackson and assured him that, while nothing had happened to *Fram*, he and Hjalmar Johansen (who was not far behind Nansen) had left the ship in March 1895 after reaching 84°N, a new Farthest North. As currents were by then dragging *Fram* away from the North Pole, they had travelled over the ice to 86° 15'N (a new Farthest North) where, on 7 April, they turned south and headed for Franz Josef Land. Nansen had Leigh Smith's and others' charts with him, but it was Leigh Smith's charts which had brought him and Johansen safely to Cape Flora.

Back at Elmwood, Jackson and Nansen reenacted their 'Dr Livingstone and Henry Stanley' encounter for the camera, so Jackson could send photographs to Harmsworth. As the Norwegians had been living rough for months, Jackson ceded his room to Nansen and suggested they might like to travel south on *Windward*, which was due at Cape Flora in July.

During his stay at Cape Flora, Nansen suggested to Jackson that, should Antarctica prove to be a large continent surrounded by sea, it might prove easier to reach the South Pole than its northern counterpart. As to transport in the south, Jackson wondered whether hardy Siberian dogs and ponies might, as they had for him here, do some of the heavy work. But this was all in the future, as while Nansen was keen to try for the South Pole, he would have expedition reports to write as well as continuing his campaign for Norwegian independence.

Windward returned to Cape Flora in late July, with provisions and letters for everyone. Armitage sadly learned that his mother had died, and for Jackson there was disappointment when he learned that under *Windward*'s charter conditions he could not sail her any further north.

Blomkvist, who was suffering from anaemia, would return on *Windward*, to be replaced by Dr William Bruce, a Scottish natural scientist with Antarctic experience who had already agreed to join Jackson for the forthcoming season.[3] When *Windward* left Cape Flora in early August, Nansen and Johansen were aboard, as were Jackson's latest reports for Markham and Harmsworth and the photographs of himself and Nansen which, Jackson hoped, would assure Harmsworth his investment in the expedition was worthwhile.

In September 1897, following a third season in Franz Josef Land, Jackson and his Cape Flora party returned to Britain, where the photograph of Jackson with Nansen had appeared on the front page of the *Illustrated London News* the previous September.[4] For Jackson, it had been a pleasure to assist Nansen, but he was nonetheless gratified when his assistance to Norway's most famous explorer was recognised more formally by a knighthood of Norway's Order of St Olaf.

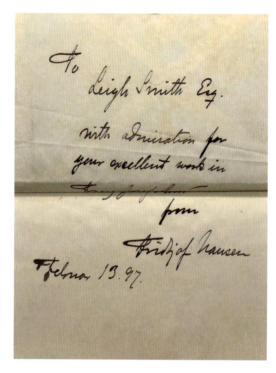

Fridtjof Nansen's dedication and thanks to Benjamin Leigh Smith, 13 February 1897, in a copy of Nansen's *Farthest North*; image © and courtesy of Orsi Libri, Milan, Italy.

POLAR POSTSCRIPT: Jackson's expedition account, *A Thousand Days in the Arctic*, sold well, as did *Farthest North*, an English translation of Nansen's expedition account. Nansen was aware of how much he owed to Leigh Smith's charts and sent the now-retired explorer a copy of *Farthest North*, in which Nansen acknowledged his fellow explorer's 'excellent [charting] work in Franz Josef Land'.[5]

Part V

Antarctica Revealed

In May 1895, Clements Markham, by now President of the Royal Geographical Society, oversaw the fiftieth anniversary commemorations of the departure of John Franklin's final Arctic expedition. One of the main events was a public exhibition of artefacts found by Leopold McClintock and others during their searches; there was also a more solemn gathering attended by McClintock and others more closely associated with the expedition and its aftermath.

Two months later, Markham – in his capacity as President of the Sixth International Geographical Congress – welcomed geographers and other delegates from all over the world to London. Among the resolutions passed was one confirming that 'exploration of the Antarctic Regions is the greatest piece of geographical exploration still to be undertaken'. Delegates also agreed that 'this work should be undertaken before the close of the century', a deadline Markham and oceanographer Sir John Murray of the *Challenger* expedition and others hoped to meet.[1] As Markham began fundraising for a two-ship expedition which would undertake both geographic exploration and scientific research, delegates, scientists and explorers from other nations worked on their own plans.

23

A Stereoview of Adrien de Gerlache and a Weddell Seal

This stereoview shows Adrien de Gerlache, commander of Belgium's 1898–89 Antarctic expedition, with a Weddell seal. The photographer is not credited but was probably Dr Frederick Cook, the expedition's doctor, scientist and semi-official photographer. Photographs taken by Cook and others during the expedition were used to illustrate expedition reports and lectures. This and related images were also sold or licensed, including to Keystone View (est. 1892), a major American distributor of stereoviews and lantern slides.

This image, together with ones of Norwegian expedition member Roald Amundsen at the ice edge and of expedition members 'watering' expedition ship *Belgica*, were marketed as a set of three by Keystone View. They were advertised until around 1913 as being available as stereoviews or lantern slides for home and educational use.

In August 1897, *Belgica*, a converted Norwegian whaler, was cheered away from Antwerp on Belgium's first Antarctic expedition.[1] Expedition commander Adrien de Gerlache was supported by a Belgian first officer, Georges Lecointe, and several Belgian scientists and crew members, but had also recruited a Norwegian second officer (Roald Amundsen), an American ship's doctor-cum-scientist (Frederick Cook) and several Norwegian crew members with previous Arctic experience.[2]

De Gerlache's instructions, which had been expressed in somewhat general terms, were to carry out geographical and scientific surveys of Antarctic coasts and to attempt to locate the South Magnetic Pole.[3] The latter involved *Belgica* depositing de Gerlache, Amundsen and a few others in northern Victoria Land (probably in early 1898), where they

A Stereoview of Adrien de Gerlache and a Weddell Seal 123

Stereoview of de Gerlache and a Weddell seal; author's collection.

Stereoview showing Amundsen at the ice edge; author's collection.

Lantern slide showing men collecting ice while 'watering ship'; author's collection.

would overwinter, take magnetic readings and then try to locate the Magnetic Pole during the following Antarctic summer. But following a series of delays, not least a major storm during which a crew member died, de Gerlache realised he had no chance of reaching Victoria Land or Melbourne by early 1898.[4] This left him with two options: return to Tierra del Fuego to overwinter or remain south and continue his surveys.

By March 1898, *Belgica* had battled through pack ice and had become the first Belgian-registered vessel to cross the Antarctic Circle and pass James Cook's Farthest South of 71° 10'S.[5] De Gerlache knew James Ross had broken through pack ice into open water, but after *Belgica* became completely ice-beset and began drifting with the currents, he ordered his men to check the provisions, re-water the ship, cover *Belgica*'s decks with awnings and prepare to overwinter.

Frederick Cook and Amundsen knew from medical training and Arctic experience that eating fresh meat helped prevent scurvy, but de Gerlache's ill-disguised distaste for both seal and penguin meat deterred others from following Cook and Amundsen's example.[6]

In early June, army lieutenant Emile Danco, a long-standing friend of de Gerlache, died of heart disease. Cook had known that Danco had an underlying heart defect, but now had to deal with men who became anxious when their hearts beat faster due to the cold, as they pumped blood round their bodies.[7] When Amundsen, probably the fittest man on board, felt his heart racing, he followed Cook's advice and huddled in front of the ship's stove to raise his core temperature. On 20 July, during one of Amundsen's regular conversations with de Gerlache regarding their predicament, his leader admitted that his sense of patriotic duty had made him continue south (against Danco and Lecointe's advice) and prioritise the achievement of Antarctic 'firsts' for Belgium over safeguarding his men's lives and health.

As the sun returned from below the horizon, the men's physical and mental well-being improved, despite *Belgica* remaining trapped in ice. Amundsen was concerned, however, that de Gerlache, who was prone to 'polar lethargy' and constant headaches, had no contingency plans and that, should anything happen to him or Lecointe, authority devolved to junior Belgian officers rather than Amundsen or Cook.

Christmas was a sombre affair, so on New Year's Eve Amundsen produced a bottle of cognac with which those not confined to their quarters could welcome 1899 with a toast, accordion music and a sing-song.

During January, as the days grew shorter again, men sawed, detonated and torpedoed the ice around the ship. As hard-won channels through the ice refroze, de Gerlache suggested they might need to evacuate and try to escape across the ice or in auxiliary boats. But on 11 February, before de Gerlache's plan could be put into action, *Belgica* began rocking gently within the huge ice floe in which she had been imprisoned for months.

In late March, *Belgica* returned to Punta Arenas, from where de Gerlache dispatched cables to Belgium, Amundsen booked a passage on a northbound steamer and Cook and Lecointe made plans to explore Tierra del Fuego. As de Gerlache, Lecointe and Amundsen bade Cook farewell, they knew how much they owed the warm-hearted American doctor who had, despite de Gerlache's initial resistance, persuaded them that Weddell seals were not only an interesting subject for scientific study, but were vital in terms of keeping men alive and healthy.

POLAR POSTSCRIPT: The photograph of Amundsen at the ice edge was published in George Lecointe's *Belgica* expedition account, *Les Pays des Manchots* (*In the Land of Penguins*). In 1912–13, Keystone View updated the caption to reflect Amundsen's subsequent polar achievements.

24

Louis Bernacchi's Cape Adare Home

During the *Southern Cross* expedition (1898–1900) Anglo-Norwegian explorer Carsten Borchgrevink, physicist Louis Bernacchi and eight companions lived in a 15-square-foot prefabricated pine hut on Cape Adare, Victoria Land.[1] Bernacchi had been accepted for the second season of Adrien de Gerlache's Antarctic expedition, but when *Belgica* failed to reach Melbourne, he contacted Borchgrevink, who had previously visited Cape Adare during Henryk Bull's 1893–95 Antarctic expedition and was planning his own expedition to Victoria Land.[2]

The huts and adjacent buildings still stand at Cape Adare and are cared for by New Zealand's Antarctic Heritage Trust. The accommodation hut's communal living space was fitted with five sets of twin bunk beds, a stove and dining table; external cubicles and a second hut provided additional work and storage space.

In mid-February 1899, physicist Louis Bernacchi landed on Cape Adare, where he, Borchgrevink, medical officer Herlof Klövstad, zoologist Nicolai Hanson, scientific assistants Anton Fougner and Hugh Evans, mariners William Colbeck and Kolbein Ellefson and dog handlers Persen Savio and Ole Must would share a 15-square-foot hut for up to a year.

Two weeks later, *Southern Cross* left for New Zealand. As seals and penguins were also now heading north to avoid the winter weather, men went hunting to stock up the hut's larder with fresh meat. When wintry gales and blizzards set in, outdoor duties were curtailed, but when off-duty, Bernacchi was happy to read or join others for post-dinner walks on moonlit nights or watch as waves of Aurora Australis filled the skies.

Cape Adare hut exterior, 2011; image © A. Strathie.

Cape Adare hut being built; photograph from William Colbeck collection (ref. c053710026), Mitchell Library, State Library of New South Wales, image courtesy of SLNSW.

On 17 May, Norway's national day, Klövstad organised a special dinner, which culminated in numerous toasts, a rocket display and a somewhat perilous torchlight procession along the beach.[3] Bernacchi liked and admired most of his companions, although he found Borchgrevink's sometimes erratic behaviour disruptive and in stark contrast with the apparently calm professionalism of polar veterans like Leopold McClintock, whose Franklin search expedition narrative he was enjoying.[4]

As the days lengthened, temperatures rose and the new sledging season got under way, most men's moods and health soon improved, but Dr Klövstad was increasingly worried about Nicolai Hanson, who had been ill over the winter and now seemed to be deteriorating. Klövstad could identify no root cause of Hanson's ailments which, by early October, left the 28-year-old zoologist bedbound and in such distress that Bernacchi and others moved into tents to give him and Klövstad more privacy.[5] On 14 October, Hanson, following discussions with Klövstad, asked that Bernacchi and others be allowed indoors so he could bid them farewell. Evans, Hanson's assistant, who was watching out for penguins returning to Cape Adare, spotted a lone Adélie and brought it in to show Hanson. Hanson roused sufficiently to ask Evans to show him the penguin's tail and, after pronouncing the bird to be fully grown, fell back on his pillow and, soon afterwards, slipped peacefully away.

By 20 October, the day of Hanson's funeral, Adélie penguins had returned in their thousands. The empty bunk in the hut was a constant reminder of Hanson, but the busy, chattering Adélies were a welcome diversion, particularly for Evans, who was now recording their rituals and breeding cycle. The expedition's main sponsor, publisher George Newnes, had provided Borchgrevink with a Newman & Guardia kinematograph, but hopes of returning to Britain with the first moving images of penguins had already been thwarted by a faulty film magazine, leaving Bernacchi dependent on his quarter- and half-plate cameras.

As the sun rose to its zenith, men began suffering from sun blindness, so began sleeping during the day and working outdoors by the midnight sun. When open water appeared near the cape, Bernacchi began to hope that *Southern Cross* might return early, but Christmas and New Year's Day 1900 came and went with no sign of sails or masts on the horizon. As the weeks passed, Bernacchi began to suspect that they might, as he put it, be 'doomed' to remain for another year.

Early on 28 January, while most men were still asleep, there was a knock on the door of the hut and a loud shout of 'Post!' Captain Jensen of

Interior of hut, 2011; image © A. Strathie.

Southern Cross, realising that his ship's mooring was not visible from the hut, had come ashore with everyone's mail – which to everyone's relief, contained no worrying news of loved ones. Four days later, Bernacchi's 'polar incarceration' ended with a final visit to Hanson's grave and the closing of the door of the hut in which he had spent what might prove to be the strangest and most difficult year of his life.

Borchgrevink still felt he had unfinished business, so *Southern Cross* headed south along the coast of Victoria Land, where Bernacchi and others took magnetic readings which confirmed that the South Magnetic Pole had moved north-west since James Ross located it. At McMurdo Sound, Mounts Erebus and Terror were unfortunately capped by clouds, but they continued along the Great Ice Barrier until they passed James Ross's Farthest South. The Barrier had, it seemed, receded south, so they dropped anchor at a low point and, by the end of the day, made the first recorded landing on the Barrier and, on surprisingly smooth ice, sledged to 78° 50'S, a new Farthest South.[6]

When *Southern Cross* reached Hobart, Bernacchi took a fast steamer to London where, on 25 June, Borchgrevink read a paper on the expedition to Royal Geographical Society members. But while the fact that Borchgrevink's

Statue of Louis Bernacchi with camera and his sledge dog Joe (who accompanied him on a second Antarctic expedition), Hobart waterfront; image © and courtesy of Julia Fortes.

men had survived a winter at Cape Adare was clearly significant for future expeditions, Clements Markham now seemed more interested in the forthcoming RGS-sponsored British National Antarctic Expedition than in following up on Bernacchi's suggestion that Antarctica might prove to be 'another Klondike'.[7] The Royal Society, however, employed Bernacchi to write up his magnetic and other observations.

There had been times during the expedition when Bernacchi had hoped never to see the Cape Adare accommodation hut again – but he had found his scientific work so interesting that when Markham suggested he apply for a place on the British National Antarctic Expedition, he barely hesitated.

POLAR POSTSCRIPT: Borchgrevink received little official recognition for his achievements, which were overshadowed by those of subsequent expeditions. Although Bernacchi had not always seen eye to eye with Borchgrevink, he joined a campaign in 1930 which resulted in Borchgrevink being awarded the Royal Geographical Society's Patron's Medal for his 'pioneering expedition', which had been first 'to winter in the Antarctic, to travel on the Ross Barrier and to obtain proof of [the Barrier's] recession'.

25

RRS *Discovery*

RRS (Royal Research Ship) *Discovery* was the first British vessel specifically designed for polar exploration and scientific work. She was commissioned for the British National Antarctic Expedition, built in Dundee and combined the features of whaling ships with dedicated space and equipment for scientific research. She could accommodate up to fifty men, including six officers and five scientists.

As designed, *Discovery* was 172ft long, with three masts, a coal-fired engine, a heavily reinforced bow and hull and a retractable rudder (to minimise damage by ice).[1] She was built from wood (to minimise the distortion of magnetic readings) and fitted with on-board laboratories and high-specification sounding and dredging equipment. *Discovery* returned to Dundee in 1984, where she can now be seen at the city's Discovery Centre.

In early summer 1900, Lieutenant Robert Scott RN was appointed captain of *Discovery* and the British National Antarctic Expedition. Although he lacked polar experience, Scott's former commanding officer, George Egerton, had served with George Nares and Albert Markham. RGS President Clements Markham also ensured that Scott received advice from other polar veterans, including Leopold McClintock and Fridtjof Nansen.[2]

As *Discovery*'s captain, Scott was supported by three naval officers, Charles Royds, Reginald Skelton and Michael Barne, and two mercantile officers, Albert Armitage (of Frederick Jackson's expedition) and Ernest Shackleton.[3] Expedition scientists Reginald Koettlitz and Louis Bernacchi both had prior polar experience, while Edward Wilson and Thomas Hodgson had worked on *Southern Cross*'s scientific reports and 21-year-old geologist Hartley Ferrar, a great sportsman, appeared fit for anything.

Discovery, Discovery Centre, Dundee; image © A. Strathie.

Discovery left London on 29 July 1901 and, although she would sail alone, Markham was also finalising arrangements for purchasing a Norwegian whaler, to serve as expedition relief ship *Morning*.[4] During the voyage to New Zealand, *Discovery* dealt well with pack ice encountered at 60°S, but Scott was less impressed by her tendency to roll in high seas.[5]

After the long voyage to Lyttelton, *Discovery* went into dry dock, and when she finally sailed for Antarctica she was, thanks in part to shipping agent Joseph Kinsey and Admiral Sir Lewis Beaumont of Britain's Australia Station, shipshape, well provisioned and fully manned.[6] In early January 1902, *Discovery* dropped anchor at Cape Adare, where Scott left a progress report for the relief vessel's captain at the hut that Bernacchi remembered well. Scott's men would, however, much to Bernacchi's relief, live on board *Discovery* in McMurdo Sound.

Not everything went according to plan, and after Scott found his chosen location blocked by pack ice, *Discovery* dropped anchor at Cape Crozier so Scott could erect a message post (where he would deposit final details of his winter quarters), then continued along the Great Ice Barrier.

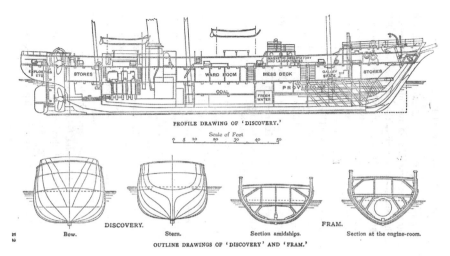

Diagram of *Discovery*, showing comparison with *Fram*; from vol. I, Scott et al., p. 51 *The Voyage of Discovery*, author's collection.

After passing James Ross's easternmost point, *Discovery* continued to the Barrier's terminal, a peninsula that Scott named for his new monarch, King Edward VII. On the way back to McMurdo Sound, they dropped anchor at *Southern Cross*'s 1900 mooring, from where Bernacchi led a sledge party south on the Barrier. Scott made the first balloon ascent from Antarctica, and Shackleton, during his balloon ascent, took the first aerial photographs of the Barrier.[7]

By early February, *Discovery* was anchored at the southern end of McMurdo Sound, within sledging distance of Cape Crozier, Mounts Erebus and Terror, the Barrier, and south Victoria Land. As working and leisure space on *Discovery* was limited, a team erected a 36-square-foot prefabricated hut on an adjacent peninsula (soon imaginatively named 'Hut Point').[8] Royds agreed to lead a party to Cape Crozier to deposit Scott's note with details of their winter quarters, but when the area was engulfed by a blizzard, seaman George Vince plunged to his death over a cliff, the party became scattered and the note remained undelivered.

In late April, as the sun dropped below the horizon, the sledging season ended, but Scott organised winter duty rotas and encouraged winter pursuits, including theatre performances (on a makeshift stage in the hut), the publication of the 'South Polar Times' (editor, Ernest Shackleton) and, in June, a Christmas-style Midwinter's Day feast. Seaman Frank Wild's draughts tournament also proved popular, but as the spring sledging season approached, Scott announced a series of scientific and other

Discovery and scientific huts; vol. I, opp. p. 280, Scott et al., *The Voyage of Discovery*, author's collection.

lectures, including his own presentation on sledging and other forms of travel used in the polar regions.

Scott's orders were to explore the western mountains of Victoria Land, the Great Ice Barrier and the area around Mounts Erebus and Terror. Thankfully, in terms of manpower, expedition doctors Koettlitz and Wilson had kept scurvy at bay during the winter. After Royds and Skelton deposited details of *Discovery*'s location at the Cape Crozier message post, Scott, Wilson and Shackleton began preparing for their journey south along the Great Ice Barrier. Wilson was fascinated by Skelton's photographs of fluffy emperor chicks at Cape Crozier but knew he would be unable to visit the penguin's breeding ground until he, Scott and Shackleton returned from their southern journey in a few months' time.

Between early November and late December, as sledging parties explored Antarctica, *Discovery* was the hub to which they returned from long sledging journeys for much-needed hot baths, fresh clothes and hearty meals. In late January 1903, as sledgers exchanged travellers' tales, a look-out saw signal rockets in the sky. Within hours, Skelton, Bernacchi and Hodgson were heading across the ice where Bernacchi's erstwhile *Southern Cross* shipmate William Colbeck, in his new role as captain of relief ship *Morning*, handed over mail from home. When the 'postmen' returned to *Discovery* and learned that Koettlitz had seen no sign of Scott's party when replenishing their Barrier food depots, they headed south to relay the news of *Morning*'s arrival to Scott, Wilson and Shackleton.

On 3 February, Skelton, Bernacchi and Hodgson finally met up with Scott, Wilson and Shackleton, who were all clearly exhausted and malnourished. Shackleton had apparently suffered most during their long journey, particularly after the dogs had died and the three men had been forced to relay sledges for hundreds of miles.[9] Scott had hoped to reach 85°S, but on 30 December had called a halt just south of 82°S.

Back at what Scott referred to as his 'beloved ship', he, Wilson and Shackleton tucked into fresh produce brought south by Colbeck and anything edible put in front of them. But although the trio's scurvy symptoms and Wilson's snow-blindness abated, Shackleton's coughing fits and breathlessness persisted.[10]

When it became evident that sawing ice and detonating explosives was unlikely to disperse the miles of ice between *Discovery* and *Morning*, Scott and Colbeck agreed that *Morning* would leave for New Zealand now, leaving *Discovery* to follow or – should the ice remain intractable – overwinter again. As there was a significant risk that *Discovery* would remain frozen in, Scott, after consulting with Koettlitz and Wilson, concluded that Shackleton should return north with *Morning* rather than risk further damaging his health during the winter or subsequent sledging season.[11]

On 1 March, after Shackleton boarded *Morning* and Lieutenant George Mulock transferred to *Discovery*, Scott and his men watched *Morning* head north. Although Scott had suggested to Colbeck and those on *Morning* that he would soon be following them to New Zealand, it was not long before he announced to his shipmates that they should begin preparing *Discovery* for a second winter in McMurdo Sound.

POLAR POSTSCRIPT: Colbeck's second officer on *Morning*, Edward 'Teddy' Evans, had, like Scott, Skelton, and seamen Evans, Quartley and Allan, served on HMS *Majestic*, and after reading Markham's appeal, had volunteered to join *Morning*. Evans was looking forward to returning to Lyttelton as he was wooing a young lady he had met there and would be happy to remain in case *Morning* needed to return south.

26

Edward Wilson's Portable Paintbox

This portable paintbox belonged to Dr Edward Wilson, the *Discovery* expedition's assistant surgeon, zoologist and artist. Wilson, who displayed an early talent for drawing, wanted to be a naturalist from the age of 9. Before joining Scott's expedition, he worked on the *Southern Cross* expedition's natural history reports and produced watercolours of seals and penguins based on skins and specimens. Wilson always bought the best equipment and materials he could afford, including this portable paintbox made by Reeves & Sons, which had a sliding lid and spaces for the artist's personal choice of blocks and half-blocks of solid watercolour paint.

The paintbox (ref. 1959.42) and a large collection of Wilson's watercolours are held in the Wilson Family Collection at the Wilson Art Gallery & Museum, Cheltenham.

During the long southern journey in the Antarctic summer of 1902–03, Edward Wilson would sit outside the three-man tent and sketch huge uncharted mountain ranges. He regularly suffered from painful snow-blindness and frost-numbed fingers, and had discovered that lead in soft B pencils froze and became as hard and gritty as those in H or HH pencils.[1] Back at the ship, Wilson wrote up his zoological reports, worked up sketches he had made on the march into finished works and kept a lookout for seals and penguins he could sketch before they left McMurdo sound for warmer climes.

As the days shortened, Wilson tried to hold in his mind's eye the colours of particularly beautiful pre-winter sunrises and sunsets. While he had always painted sunsets and sunrises, it was more of a challenge to

Edward Wilson's portable paintbox; image © and courtesy of The Wilson Family Collection at Cheltenham Borough Council/The Cheltenham Trust.

Earth shadows (showing subtle shades and celestial phenomena achievable in paints, rather than photographs); vol. II, opp. p. 196, Scott et al., *The Voyage of the Discovery*, author's collection.

capture the essence of Aurora Australis and other celestial phenomena. He had also realised that icebergs which looked white on first glance reflected the changing colours of the sea and sky and were run through with shadowy crevasses and veins of bright cobalt blue or aquamarine.

As Wilson worked his outdoor sketches into finished watercolour paintings, he relied (even more than usual) on his near-perfect memory for colours. He would jot notes on his initial sketches: 'palest blue-green to lemon yellow' (for a sky); 'tinge of blue, deep pure grey' (a mountain range); 'grey shadows and footmarks' (a foreground). After mixing the required colours from his blocks of paint, he selected brushes from his large collection which ranged from small ones he used to paint the feathers of adult birds or down on a chick, to larger, softer ones, suitable for applying translucent washes across large areas of sky.

Wilson's work was much in demand, including from Scott (for his expedition report) and from Louis Bernacchi who, following the departure of 'Shackles', now edited the 'South Polar Times'. Wilson was one of the periodical's most regular contributors, whether of sketches of sledge flags, cartoon-style drawings or more serious fare including illustrated articles on seabirds and penguins.

Wilson freely admitted that he knew little about the annual breeding cycle of the emperor penguin, which no scientists had studied in detail. Based on photographs Reginald Skelton had brought back from the emperor rookery at Cape Crozier and a questionnaire Wilson had given him and Charles Royds to complete, Wilson now wondered whether emperor penguins laid and incubated eggs during the Antarctic winter. Should that be the case, once chicks were part-grown, emperors might migrate north like other penguin species. It was a mystery, but with *Discovery* frozen in until spring and *Morning* or other relief ship unlikely to reach McMurdo Sound until early 1904, Wilson now hoped to make several sledge trips to Cape Crozier.

While Wilson planned his zoological campaign, Scott revised his plans for the next sledging season. If *Discovery* had not remained frozen in for a second winter, he could have explored King Edward VII Land then returned to Britain via Cape Horn. That plan and his hopes of locating the South Magnetic Pole now appeared to be hostages to fortune, but after reviewing the previous season's sledge journeys in detail, Scott realised that, based on what they had already achieved, they could ascend glaciers terminating on the Great Ice Barrier and try to reach the polar ice-cap and find a route to the South Geographic Pole.

Cape Crozier emperor penguin colony, watercolour by Edward Wilson; vol. II, opp. p. 212, Scott et al., *The Voyage of the Discovery*, author's collection.

In early September, as daylight returned, Wilson and Royds led a six-man party to Cape Crozier, where they found about 1,000 emperor penguins, some nursing chicks noticeably larger than the ones Skelton had photographed later in the season the previous year. Wilson wondered whether emperor chicks, which the parents kept warm in a pouch near their feet, might be born during the winter, and after finding some eggs and frozen chicks, he took them back to the ship, where he catalogued, sketched and made paintings of his specimens.

On a return visit, Wilson deposited an updated report from Scott in the message post. There were no eggs yet in the Adélie colony, and he noticed the emperor chicks were larger, but as they were still down-covered, he knew they would not survive in the water. During Wilson's third visit in November (a year since Royds had visited the rookery and found no chicks in evidence), he watched in amazement from Mount Terror's blizzard-swept lower slopes as huge rafts of fast ice broke away from the shore and, carrying hundreds of down-covered emperor chicks and a few 'guardian' adult birds, began drifting north. Back at the ship, Wilson got out his paints and began illustrating what he regarded as his most interesting zoological discoveries to date.

Scott was also pleased with the season's work, not least with a marathon two-month journey during which he and seamen Tom Crean and Edgar Evans had, in bitterly cold weather, scaled glaciers and explored the Antarctic plateau – and on several occasions, nearly come to grief as they tumbled into crevasses and down ice falls when still roped to each other or their sledge.

When *Morning* arrived in early January 1904, she was accompanied by *Terra Nova*, a Dundee whaler chartered by the Admiralty in case *Discovery* had to be abandoned. *Discovery* was still frozen in fast, but on 16 February, after weeks of hard effort, the miles-wide belt of ice between her and open water was breached and she finally floated free.

After returning to Britain, Wilson worked on illustrations for Scott's two-volume expedition report, *The Voyage of the Discovery*. When published, it included over 250 illustrations which contained thirteen coloured plates and a photogravure frontispiece based on his watercolours, and large numbers of his scientific and other drawings. A selection of his paintings and drawings were exhibited alongside Skelton's photographs (which also featured in the report) at Bruton Gallery in New Bond Street, London. Wilson's paintings were much admired, and reviewers and others remarked that those of the penguins and other wildlife looked 'alive' in comparison to illustrations based on skins or stuffed specimens. Exhibition visitors could order copies of Skelton's photographs and Wilson's watercolours. While replicating Skelton's images was straightforward, Wilson found himself spending much of 1904–05 painting near-replicas of his images of penguins and other popular subjects.[2] It was wearying work, but Wilson and Scott both knew that, even in the so-called age of photography, the paintbox and brush still had important roles to play in communicating the grandeur, beauty and wonders of Antarctica.

POLAR POSTSCRIPT: Wilson was particularly gratified that 82-year-old naturalist Joseph Hooker, who had served as junior naturalist on HMS *Erebus* six decades ago, visited the exhibition. Hooker made his admiration for Wilson's work widely known and invited Wilson to visit him so that he could view Hooker's Antarctic portfolio and reminisce about what they had seen, done, drawn and painted.

27

A Postcard of Three Scottish Scientists

Three scientists (left to right), Robert Rudmose Brown (botanist), David Wilton (zoologist) and James Pirie (geologist), were photographed during William Bruce's 1902–04 Scottish National Antarctic Expedition wearing traditional patterned jerseys made in Fair Isle, one of Scotland's Shetland Islands.[1] The patterns on the jerseys and Fair Isle knitted gloves, scarves and headgear are decorative, easy for the owner to identify and have an additional layer of wool. The Fair Isle knitwear was donated by the expedition's main sponsors, James and Andrew Coats, directors of a Scottish-based but worldwide family textiles business.[2] Bruce met the Coats brothers in the late 1890s, when he was serving as scientist on Andrew Coats's Arctic expedition on the latter's steam yacht *Blencathra*.[3]

The postcard, from a set of twelve *Scotia* expedition postcards, was published by William Ritchie & Sons of Edinburgh and London.

On 2 November 1902, William Bruce's expedition ship *Scotia* was piped and cheered away from Troon Harbour in Ayrshire.[4] Bruce was a proud Scot and, although *Scotia* was Norwegian-built, she had been refurbished and fitted with laboratories in an Ayrshire shipyard and her crew and other expedition personnel were almost all Scottish. Bruce was, however, internationally minded and was in regular contact with scientists and explorers from other countries, including Germany's Erich Drygalski and Sweden's Otto Nordenskjöld, with whom he might cross paths during his forthcoming expedition.

In January 1903, *Scotia* docked in Stanley in the Falkland Islands, where Bruce had arranged to set up a meteorological station. As they continued

Scotia expedition postcard no. 8, showing scientists in the on-board laboratory (see text); author's collection.

Fair Isle jersey, *c.*1912, showing colour range of natural dyes (ref. TEX81320); image © and courtesy of Shetland Museum and Archives, Lerwick.

south, temperatures dropped and Bruce issued additional winter gear: notably brightly coloured Fair Isle jerseys, mitts, gloves and scarves which expedition sponsor James Coats had commissioned from Fair Isle specially for the expedition. They encountered pack ice at around 60°S, but once south of the South Orkneys they re-entered open water and had a clear run to the Weddell Sea, where they became blocked again at around 70°S.[5]

Rather than risk becoming frozen in, they returned to the South Orkneys, where *Scotia* dropped anchor off a sheltered bay on Laurie Island. As Bruce had not brought expensive prefabricated wood panels, the men raided the carpenter's store, then began collecting rocks and other materials so they could build Omond House, which Bruce hoped might become the first permanent scientific station in Antarctica.[6]

Most of Bruce's men were accustomed to 'dreich' northern winters and eating fresh fish for breakfast and were willing to eat penguin and seal meat (preferably cooked with onions) if that kept scurvy at bay.[7] In terms of exercise, Bruce, a keen skier, encouraged men to join 'ski-running' outings to a nearby glacier, down which he and other experienced skiers sped at up to 40mph.[8] The men also enjoyed singing, bagpipes and other music, and occasional special feasts, including a Midwinter's Day dinner, following which men enjoyed the contents of a cask donated by Messrs Guinness of Dublin, which turned out to be part-frozen and had an unusually high alcohol content![9]

Bruce was usually busy with scientific and other work, but he was concerned for his men's welfare during the dark months, so he regularly covered night watches and tutored 'Shetland Johnnie', the youngest crew member, in arithmetic and reading. But Bruce and Pirie, despite their medical training, could do nothing to help Chief Engineer Allan Ramsey, who died in August from a previously undiagnosed heart condition and was buried far from home to the sound of Gilbert Kerr's pipes playing the Scottish lament 'Flowers of the Forest'.

As the men explored the island overland or in boats, they took with them Arctic-style fur coats, which kept them warm at night. But as furs were too bulky and hot to work in, they usually wore their Fair Isle jerseys and tweed trousers, with woollen underwear and a waterproof outer layer, which covered most eventualities.[10] As their extremities were at risk of being frost-nipped, they also regularly wore Fair Isle hats, scarves and gloves (including indispensable fingerless mitts) and an ingenious Jaeger soft woollen cap with a brim which concealed drop-down panels to protect the neck and face.

'The Piper [Gilbert Kerr] and the Penguin'; opp. p. 236, Pirie et al., *The Voyage of the Scotia*, author's collection.

When the ice released *Scotia* in November, Bruce decided to return to Buenos Aires, where he would submit a request to British officials for future funding for Omond House, while Pirie, zoologist Robert Mossman and a small group remained at Omond House itself. In Buenos Aires, Bruce learned that Drygalski's *Gauss*, Nordenskjöld's *Antarctic* and Scott's *Discovery* had all been ice-beset for at least a year. He realised he had been relatively fortunate but was disappointed when British officials in London declined to support the continuation of Omond House as a scientific station.[11] They did, however, authorise Bruce to transfer Omond House, its ancillary buildings and specified equipment to Argentina's Meteorological Office if they were prepared to operate it from Buenos Aires. Following meetings with Argentinian officials, *Scotia* returned to Laurie Island in February 1904 with three Argentinian scientists, who would join Mossman and Smith until the remainder of the Argentinian team arrived.

When *Scotia* returned to the south-eastern Weddell Sea she reached 74°S, from where Bruce and others saw a land mass which, judging by soundings, was probably part of an Antarctic continent. Thick pack ice precluded landings, but Bruce charted the land mass as Coats Land in tribute to his generous supporters. While *Scotia* was stuck in the ice, the men charted an ice barrier (of similar height to that charted by Ross) and established that emperor penguins were not particularly fond of Gilbert Kerr's bagpipe playing.

In July 1904, after almost two years away, *Scotia* returned to the Firth of Clyde. The Coats brothers met her in their yachts and escorted them to Millport, where Bruce and his team were cheered ashore. After Sir John Murray read out King Edward's telegram of congratulations, he presented Bruce with the Royal Scottish Geographical Society's Gold Medal. Bruce's men had made extensive oceanographic surveys, brought back crate-loads of geological and other specimens, recorded over 1,000 species (including over twenty new ones), charted new land, kept men scurvy-free, secured the future of Omond House and made what might be the first moving images of icebergs and penguins.[12]

Back in Edinburgh, Bruce began writing up scientific reports and lobbying for the establishment of a permanent Scottish oceanographic laboratory. As he had little free time, he delegated the writing of an expedition narrative to Pirie, Rudmose Brown and Mossman, who produced *The Voyage of the 'Scotia'* under the collective pseudonym 'Three of the Staff'. The book was dedicated to Bruce, whom the authors described as their 'comrade and leader'.

There were no frontispiece photographs of the three co-authors, but those who knew them would recognise them and other scientists, wearing their distinctive Fair Isle jerseys, working in *Scotia*'s laboratory, on deck or ashore. For them and for Bruce, science was a collective endeavour; although, as Bruce explained in his brief introduction to *The Voyage of the 'Scotia'*:

> While 'Science' was the talisman of the Expedition, 'Scotland' was emblazoned on its flag; and it may be that, in endeavouring to serve humanity by adding another link to the golden chain of science, we have shown that the nationality of Scotland is a power that must be reckoned with.

POLAR POSTSCRIPT: In 1909, Bruce presented a signed copy of *The Voyage of the 'Scotia'* to *Discovery* expedition naturalist Edward Wilson.[13] Wilson and his *Discovery* colleagues were, on returning from Antarctica, presented with a new-style Polar Medal (a successor to the Arctic Medal). As this was limited to government-approved expeditions, Bruce and his men were not eligible for the medal, but in 1910 Bruce was awarded the Royal Geographical Society's Patron's Medal 'for explorations in the Arctic and Antarctic'.[14]

28

ARA *Uruguay*

Between 1903 and 1905, *Uruguay*, a 30-year-old Argentinian ex-gunboat, single-handedly relieved and assisted three Antarctic expeditions.[1] *Uruguay* was built in 1874 by Laird Brothers of Birkenhead, near Liverpool. After service as a gunboat, training ship, naval headquarters and scientific observation vessel, she was refitted for expedition support and in 1903, following an increase in the number of expeditions to Antarctic regions, was fitted out as a relief and rescue vessel.

Uruguay remained in active service until 1926 and was used by the Argentinian navy until 1962. She now serves as a floating museum in the Puerto Madero area of Buenos Aires.

Before Swedish explorer Otto Nordenskjöld's expedition ship *Antarctic* left Buenos Aires in late 1901, Argentinian naval officer José Sobral joined the expedition's scientific team. In early 1902, *Antarctic*'s captain, Carl Larsen, dropped Nordenskjöld, Sobral and a group of scientists on Snow Hill Island, off eastern Graham Land, where they would overwinter. Larsen would return north for surveying work and return in late 1902 or early 1903 to collect Nordenskjöld's party and bring Sobral back to Buenos Aires.

When *Antarctic* failed to return to Buenos Aires as planned, the Swedish authorities dispatched the whaler *Frithjof* to relieve Nordenskjöld and his men.[2] But as *Frithjof* had 8,000 miles to travel, the Argentinian authorities decided to deploy *Uruguay*, its own relief and rescue vessel. When Lieutenant Commander Julián Irízar, Argentina's naval attaché to Britain, was given command of the rescue expedition, he travelled to Dundee and Norway to seek advice and procure additional equipment, and conferred with Ernest Shackleton, who was at that time working at the Admiralty.[3]

ASA *Uruguay*, now a museum, moored in Puerto Madero, Buenos Aires, 2023; image © A. Strathie.

Irízar was naturally keen to succeed in his first polar commission but was also, on a personal basis, concerned for José Sobral, who had served under him as a junior officer.[4]

In early October, as he prepared to leave Buenos Aires, Irízar learned that *Frithjof* had reached Madeira and that French explorer Jean-Baptiste Charcot was also on his way south and willing to postpone his own Antarctic expedition to assist in the search-and-rescue operation.[5] By November, following a difficult journey south, Irízar and his men were off Graham Land, from where they headed for Seymour Island, where Irízar hoped to find a message at Nordenskjöld's main provisions depot.[6] There was no message, but his landing party found a makeshift cairn and a boathook handle carved with the words 'Sobral, Andersson, October 1903'. Irízar was baffled as to what Sobral (based at Snow Hill) and Andersson (an *Antarctic* crew member) meant by leaving the boathook at the message cairn. Although Irízar had not seen *Antarctic* on the way south, it could only mean that she had already collected the Snow Hill

The hut on Snow Hill Island, where Nordenskjöld's party overwintered; image © and courtesy of Eva-Linn Röjerstrand.

party or that Andersson had arrived with a message from Larsen for the Snow Hill party.

On 8 November, while Irízar searched around Snow Hill for a safe mooring for *Uruguay*, a lookout noticed what appeared to be a tent on the shore. On landing, Irízar and another officer found two members of Nordenskjöld's party, who had been sent to wait for *Antarctic*. Irízar was relieved to learn that everyone at Snow Hill was well, despite having spent a second unplanned winter there. As to why Andersson was now at Snow Hill, it seemed that after becoming ice-beset over a year ago, Larsen had landed Andersson and two other *Antarctic* crew members so they could travel to Snow Hill and tell Nordenskjöld that the ship was delayed. But as Andersson and his two companions had since overwintered and only just reached Snow Hill, they now had no idea where *Antarctic*, Larsen or their shipmates were.

After mooring *Uruguay*, Irízar and a small party rowed ashore and walked the remaining 12 miles to Snow Hill. Nordenskjöld was delighted to see Irízar and agreed to begin packing up at Snow Hill while Irízar

Bust of Julián Irízar, Ushuaia, 2023; image © A. Strathie.

returned to *Uruguay* and tried to bring her closer for loading. After Irízar found a suitable spot, Lieutenant Hermelo led a small party to Snow Hill so they could guide and assist Nordenskjöld's men. But before Hermelo returned, Irízar saw Captain Larsen of *Antarctic* approaching his ship.

It was, everyone agreed, a 'Day of Miracles'. Larsen and a party from *Antarctic* had reached Snow Hill after rowing over 100 miles and tramping for miles across ice and snow. As Larsen explained, after he had landed Andersson's party in December 1902, *Antarctic* had become frozen in and drifted with the current until, in February 1903, she had been crushed by the ice and begun sinking about 20 miles off Paulet Island. In spring Larsen and his men had managed to load and lower auxiliary boats, which they had dragged across the ice to Paulet Island. All, bar one crew member with heart disease, had survived the winter in makeshift winter quarters and on reduced rations. In spring, Larsen had led a party back to where he had left Andersson's party. There they found a note from Andersson explaining that, after overwintering, his party were now heading for Snow Hill, and they expected to find Larsen and *Antarctica* already there.

With Nordenskjöld's party and all bar one group from *Antarctic* accounted for, Irízar set course for Paulet Island, but before leaving they replenished Nordenskjöld's food depot at Seymour Island, with a view to assisting future ice-beset travellers. At Paulet Island, they collected the final group from *Antarctic* and, after replenishing their food depot, headed for Buenos Aires.

Given what Nordenskjöld and Larsen's men had suffered, Irízar hoped to reach Buenos Aires as soon as possible, but events took another turn when a storm sprung up during which *Uruguay*'s two main topmasts were so badly damaged that Irízar decided to collapse them and continue by steam. On 2 December 1903, *Uruguay* was cheered into port at Buenos Aires, where Charcot, Bruce and others were waiting to greet Nordenskjöld, Larsen and their men and to congratulate Irízar and his men on a truly remarkable search-and-relief mission.[7]

POLAR POSTSCRIPT: Irízar was promoted and decorated by Argentinian and other organisations, then returned to regular naval duties. Charcot was so impressed by Irízar's efforts that he named islands off Graham Land for both Irízar and *Uruguay*.

José Sobral so much enjoyed his time at Snow Hill that he resigned his naval commission and moved to Sweden to study geology with Nordenskjöld at Uppsala University. When he returned to Argentina, he worked as National Director of Mining and Hydrology and as a meteorologist in the whaling industry.

Uruguay was repaired and refurbished and in late 1904, now under Captain Galindez, sailed south to Laurie Island with five Argentinian scientists, who took over from Mossman and Smith at Omond House. As by then there were concerns for Charcot and *Français*, *Uruguay* continued to Deception and Weincke Islands (both on Charcot's planned itinerary), where Galindez deposited messages for Charcot in cairns.

Larssen returned to South Georgia, where he had already made surveys for Nordenskjöld and where, by 1904, he had established a major whaling operation at Grytviken.

Part VI

Striving for Polar Firsts

As the pace of exploration of Antarctica quickened, *Belgica* expedition veteran Roald Amundsen prepared to make the first continuous traverse of the Northwest Passage and establish the current location of the North Magnetic Pole. Amundsen's mentor, Fridtjof Nansen, still hoped to reach the geographical North Pole on his exploration vessel *Fram*, but as he was increasingly preoccupied with campaigning for Norway's independence from Sweden, the field was left open for American explorer Robert Peary to continue his quest to reach the pole by travelling across sea ice.

Amundsen, Nansen, Peary and other Arctic explorers continued to favour traditional equipment and means of transport used by the Inuit people, whalers and other mariners with long experience of the Arctic. But as technology advanced, others began to explore the possibility of using motorised transport in polar regions, including on Antarctica's Great Ice Barrier.

29

Amundsen's Dip Circle

Roald Amundsen used this dip circle during his traverse of the Northwest Passage from 1903 to 1906. Amundsen had, since returning from the *Belgica* expedition, obtained his seaman's certificate, studied terrestrial magnetism with Professor Georg von Neumayer in Hamburg and purchased a complete library of books on his boyhood hero John Franklin and Northwest Passage expeditions. Amundsen, like John Rae, favoured travelling with small groups of men, so when planning for his Northwest Passage expedition he recruited a small team with Arctic experience and a broad range of skills: Godfred Hansen (navigator, astronomer, geologist, photographer); Peder Ristvedt (engineer, meteorologist); Gustav Wiik (engineer, magnetologist); Anton Lund (first mate); and Adolf Lindström (cook).

The magnetic dip circle was made in Kent, England, by a company founded by John Dover (1824–81), who supplied dip circles to the Admiralty. It is now in the collection of the Fram Museum, Oslo, which also includes Amundsen's expedition ship *Gjøa*.

On the night of 16 June 1903, under the midnight sun, Roald Amundsen and his companions left Christiania, Norway's capital, on *Gjøa*, a converted 70ft fishing sloop.[1] Amundsen had, with assistance from his brothers and mentor Fridtjof Nansen, raised sufficient funds to cover the purchase and refurbishment of *Gjøa* but other suppliers were now pressing for payment.[2]

As Amundsen needed to complete his expedition before he could raise further funds from newspaper 'exclusives', an expedition narrative or lecturing, he decided to slip away quietly. After brief stops in Baffin Bay to refuel and collect equipment, sledge dogs and additional provisions, Amundsen headed for Lancaster Sound and the Northwest Passage.

On 22 August, *Gjøa* dropped anchor at Beechey Island. As Amundsen stood on the beach, he imagined his boyhood hero John Franklin landing

Amundsen's dip circle; image © and courtesy of Fram Museum, Oslo.

and considering whether to overwinter in the sheltered bay before continuing his long voyage.[3] While there, Amundsen paid his respects at the lonely graves on the foreshore and the marble slab which McClintock had erected in 1858 to Franklin and his men – who were, in Amundsen's view, 'the discoverers of the Northwest Passage'.

The following day, in foggy conditions, Amundsen and his men continued to explore the island and took their first magnetic observations. As the dip circle needle settled, it suggested that if Amundsen and Wiik wanted to locate the North Magnetic Pole, they needed to head south before continuing west.

Amundsen found it hard to leave the melancholy island. He discouraged men from bringing bulky mementos aboard, but agreed to find space for an old anvil which might prove useful.[4] Thankfully, Peel Sound and Franklin Strait were relatively ice free, but when *Gjøa* entered James

Ross Strait, the ice became heavier. Their compasses became increasingly unreliable as they approached the area of western Boothia where Ross had located the Magnetic Pole in 1831, so they resorted, like Viking explorers, to navigating by the stars. It seemed from their readings that the Magnetic Pole had moved little since James Ross had searched for it seven decades previously, but Professor Neumayer had recommended to Amundsen that he find a safe mooring about 100 miles from the Magnetic Pole, so he could take consistent readings for a full year.

As they continued south along the east coast of King William Island, a storm arose during which *Gjøa* was grounded on submerged rocks. This trial by water was succeeded by an engine room fire, but by mid-September they were safely moored in an inlet off Rae Strait and settled down to embark on their programme of magnetic and other observations.

By summer 1904, Wiik had completed a year's magnetic observations, which suggested that the pole was still on western Boothia but had moved slightly north-east of Ross's coordinates.[5] Another echo of the past came when Hansen and others, while exploring the island, found a cairn which contained what appeared to be the remains of two or three members of Franklin's expedition.[6]

By now, Amundsen increasingly left magnetic work to Wiik, and passed more of his time with members of local Inuit communities who, after coming to view the ship, set up camp nearby. Over the rest of the year and into 1905, the Inuit shared their local knowledge and traditional skills with Amundsen and his men. They seemed happy to exchange hand-embroidered and fur garments for food and items from the ship which caught their eye. Thanks to the Inuit, Amundsen and his men became increasingly adept at kayaking and hunting, so rarely lacked for fresh meat.

When the ice at 'Gjøahavn' went out in late July, Amundsen and his men began to pack up and, after making their farewells, continued west. As they followed the northern coast of Canada westwards, they enjoyed being in open water, but when they encountered thick ice near the mouth of the Mackenzie River, they prepared to overwinter at King Point on Herschel Island.

Amundsen was becoming anxious to have news from home and to send messages to his brothers, who were waiting to sell 'exclusives' to newspapers, as he had now passed through the notorious Arctic archipelago. When he learned that ice-beset American whalers were organising a mailing run to Eagle City in the Yukon in late October, he packed his skis and joined them for the 400 miles to the telegraph station.

Postcard of *Gjøa* in Golden Gate Park, San Francisco; author's collection.

Shortly after Amundsen dispatched cables and a 1,000-word report to Norway, temperatures dropped so precipitously that the telegraph wires froze and snapped. It was unclear when the wires would be repaired, but it was not until early January 1906 that Amundsen learned that family members were well, his brothers were managing to raise more money and, partly thanks to Nansen, Norway was now independent of Sweden.

Amundsen returned to King Point in early March to find that Gustav Wiik was unwell. They were far from any medical facilities, so there was little they could do for him and he died in late March. As the ground was still frozen solid, they entombed him in the makeshift magnetic observatory where he had taken his last readings.

After several false starts, Amundsen and his men continued west and in September they emerged through the Bering Strait. Their first port of call was Nome, Alaska, where members of the settlement's Norwegian community cheered them into harbour.

In San Francisco, members of the Norwegian community were glad to have cause for celebration, following the huge earthquake which had devastated their city in April. As *Gjøa* was too small and too damaged to return to Norway by way of Cape Horn, Amundsen left her in San Francisco, from where he and his men would make their own way home. Amundsen took a train and, after giving a few lectures, returned to

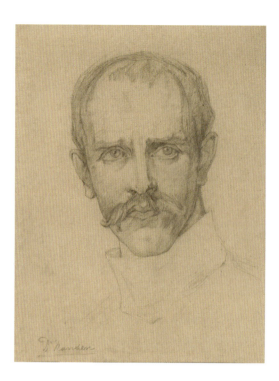

Sketch portrait by Violet Manners of Fridtjof Nansen, Amundsen's mentor, who became newly independent Norway's first ambassador to London in 1906; image © and courtesy of Allegra Huston.

Norway, where his triumphant homecoming could not have been more different from his clandestine departure three years previously.

In early 1907, Amundsen was invited to lecture to members of Britain's Royal Geographical Society, which had contributed £1,000 towards his expedition. During his talk, he made a point of paying tribute to Gustav Wiik, whose magnetic observations were the cornerstone of the expedition's scientific findings. Fridtjof Nansen, who was now Norway's first Ambassador to Britain, attended the lecture and was also present when Amundsen was awarded the society's 1907 Patron's Medal for 'his daring voyage for the purposes of research in the region of the North Magnetic Pole, and for his first accomplishment by any vessel of the famous North-West Passage'.

POLAR POSTSCRIPT: In 1909, *Gjøa* was installed in San Francisco's Golden Gate Park, where she was a popular visitor attraction for many years. She was later returned to Norway and, after refurbishment, was installed at the Fram Museum in Oslo.

30

Shackleton's Sledging Compass

This compass was one of a pair used by Ernest Shackleton during his 1907–09 Antarctic expedition, when he attempted to reach the South Pole. Shackleton had, since returning from Scott's *Discovery* expedition, struggled to find a fulfilling new role and in 1906, following the publication of Scott's *Voyage of the Discovery*, produced a document entitled 'Plans for an Antarctic Expedition to proceed to the Ross Quadrant of the Antarctic with a view to reaching the Geographical South Pole and the South Magnetic Pole'. Despite embarking on numerous business ventures, Shackleton had no personal capital but when his employer, Glasgow-based industrialist William Beardmore, offered to guarantee bank loans on his behalf, Shackleton's planned expedition became viable.[1]

The compact compass (approximately 3in by ½in when closed) was made by Francis Barker & Son and (as indicated on the compass dial) supplied by Cary, Porter & Sons of London. The aluminium compass card is luminous and the main compass points are painted in red on the underside of the glass for easy reading in low visibility. The compass is in the collection of South Georgia Museum (ref. 1997.6.259) and was donated by Shackleton's granddaughter, Alexandra Shackleton.

On 3 January 1909, after taking theodolite readings at 87° 22'S, Ernest Shackleton realised that he, Jameson Adams, Eric Marshall and Frank Wild must either abandon their attempt to reach the South Pole or risk running out of rations before returning to their expedition base in McMurdo Sound.[2] The foursome had already passed the Farthest South set during the *Discovery* expedition and reached the highest latitude, north or south, attained by any explorer. But before he turned back,

Shackleton's sledging compass, South Georgia Museum (ref. 1997.6.259); image © and courtesy of South Georgia Museum.

Shackleton wanted to plant a Union Jack within 100 geographical miles of the South Pole.

On 9 January, after two days trapped in a blizzard-bound tent, Shackleton and his companions prepared for a final march south. They were now at high altitude, so Shackleton dispensed with their heavy theodolite and sledge so they could travel on foot carrying minimal rations and equipment. After five hours' hard slog they reached what Shackleton estimated using dead reckoning to be 88° 23'S 162°E, about 97 geographical miles from the South Pole.[3]

Peering through his powerful binoculars, Shackleton saw nothing ahead bar the flat, featureless plateau he now assumed stretched to the pole. As an icy gale whipped round them, he got out his camera so he and Marshall could capture the moment for posterity. That done, they refolded their

Marshall, Wild and Shackleton at their Farthest South; from vol. I, opp. p. 348, *The Heart of the Antarctic*, author's collection.

flags, buried a brass cylinder containing written evidence of their presence and packed up their tent. They still had some 700 miles to cover before they reached their expedition base at Cape Royds where, all being well, *Nimrod* would already be in McMurdo Sound, ready to collect them and take them back to New Zealand.

Icy southerly winds and blizzards initially propelled the quartet northwards across the plateau. They managed to eke out their rations until they reached their food depot at the top of the Great Glacier – the route to the plateau that Shackleton had found by following a more easterly traverse of the Great Ice Barrier than he had with Scott.

As Shackleton forged onwards, however, he began weakening and as they left the plateau his pulse became increasingly irregular. He revived sufficiently to negotiate the crevasse-ridden ice falls of the Great Glacier but was grateful when Marshall handed out Burroughs's 'Forced March' tablets to everyone.[4] Although the tablets provided a welcome boost, they soon ran out, and after Wild and Adams collapsed, Marshall sped ahead to the next depot and returned with some high-protein rations.

In early February, as they hauled their sledges across the Barrier, blizzards concealed food depots and replaced crevasses as their main hazard. And while 'hooshes' made from the flesh of ponies which had died on their outward journey could be nourishing, if the meat was off or undercooked, debilitating attacks of diarrhoea and dysentery soon followed. They were all now so thin that they could feel the sledge harnesses through their clothes but, spurred on by hallucinatory food dreams, they ploughed on to their next food depot.

On 22 February they noticed some sledge tracks, and followed them to a small food depot where they found a tin can of a different brand from the ones they had brought from England. As that could only mean that *Nimrod* had arrived, they were spurred on again. The following day, they reached their final Barrier food depot on Minna Bluff, which their thoughtful companions had restocked with staples and with luxuries beyond their wildest food dreams. But while food was no longer an issue, they now faced a new challenge.

Shackleton had given instructions for *Nimrod* to leave McMurdo Sound on 1 March in the event they had not yet returned from the south. Although they were now on familiar territory with little need for compasses or other navigational instruments, after Marshall collapsed and blizzards swept in, it became a race against time. When Shackleton and Wild reached Hut Point, there was no sign of *Nimrod* or other expedition members, so they improvised an emergency flare of wood from their old magnetic hut, which eventually attracted the attention of the men who were loading the ship. All they had to do now was retrieve Marshall and Adams from the Barrier where they had left them.

On 9 March, *Nimrod* left McMurdo Sound and headed north. Shackleton was disappointed to have missed reaching the pole by such a short margin, but otherwise felt that all had turned out well. Edgeworth David and Douglas Mawson and their party had located the South Magnetic Pole and David, Mawson and others had also made the first recorded ascent of Mount Erebus. The expedition's ground transportation had, by and large, worked well, with ponies hauling heavy provisions across the Barrier and dog sledges and a car supplied by William Beardmore being used for short-distance transport. And thanks to his team of scientists and photographers, Shackleton would return to Britain with sufficient scientific readings and reports, photographs and moving images, and charts and maps to satisfy everyone from members of learned societies to the general public.

On 23 March, Shackleton landed on Stewart Island, south of New Zealand, from where he sent a prearranged exclusive coded cable and

2,500-word report to the *Daily Mail*'s offices in London. By the time *Nimrod* had landed in Lyttelton and Shackleton travelled to Christchurch, the story was in the newspapers and he and his men were heroes.

While *Nimrod* went into dry dock, Shackleton caught a steamer to Australia, where he learned, to his considerable annoyance, that Clements Markham, who had wished him well before *Nimrod* sailed, was querying his Farthest South coordinates on the grounds that Shackleton had used only his compass and dead reckoning to determine them. Markham also suggested that because Shackleton had, after finding King Edward VII Land inaccessible, based his expedition at McMurdo Sound, he had pre-empted Scott's plans to return south.

Markham was, however, no longer President of the Royal Geographical Society, and when Shackleton returned to Britain he lectured to the society's members at Albert Hall, where the Prince of Wales presented him with a gold RGS medal and his men with silver medals. By the end of the year, Shackleton had been awarded a £20,000 government grant towards still-unpaid expedition expenses and been knighted by the king. His expedition narrative, *The Heart of the Antarctic*, was a popular success, as were his expedition films, which Gaumont were now distributing nationwide following a royal command showing at Balmoral. Shackleton was the man of the moment but he knew he could not afford to rest on his laurels, especially after Robert Scott announced publicly that he was now planning to return south.

POLAR POSTSCRIPT: In 2009, to mark the centenary of the *Nimrod* expedition, the late Henry Worsley carried to the South Pole a near-identical *Nimrod* expedition sledging compass used by Shackleton.

31

Deception Island

French explorer Dr Jean-Baptiste Charcot was more interested in scientific exploration than claiming polar firsts. He was, however, interested in innovation, and before leaving on his second Antarctic expedition he invited his friend Robert Scott to join him in the French Alps so they could collaborate on test-driving the motorised sledges they planned to deploy in Antarctica. Charcot had, as previously, planned an extensive scientific programme, which included a survey of Deception Island – something he had hoped to carry out on his first expedition but had been unable to do due to his ship, *Le Français*, being damaged.

Deception Island lies at around 63°S 60°W. One of the South Shetland Islands, it is almost circular and covers around 30 square miles. During the 1820s it was regularly visited by sealers and whalers, but following volcanic eruptions in 1839 and 1842, it was generally only used as a harbour until shortly before Charcot's second Antarctic expedition.

In mid-August 1908, Charcot left Le Havre on his new vessel, *Pourquoi Pas?*. His wife Marguerite, whom he had married the previous year, accompanied him as far as Punta Arenas, from where Charcot and his men headed south to their first Antarctic port of call. On 22 December, as *Pourquoi Pas?* approached Deception Island, Charcot was surprised to see two small whaling vessels, one pulling a whale, heading in the same direction. The captain of one of the ships hailed Charcot and offered, in perfect English, to escort *Pourquoi Pas?* through the narrow channel leading to the island's main basin.

Charcot had been told during his previous expedition that few ships visited Deception Island, so he was surprised to find the basin so full

Deception Island

Whaler's Bay, Deception Island, with *Pourquoi Pas?* moored alongside Andresen's *Gobernador Bories*; from p. 311, *Le Pourquoi Pas? dans L'Antarctique*, 1908–10, author's collection.

Pourquoi Pas? re-coaling in Pendulum Cove; from p. 43, *Le Pourquoi Pas? dans L'Antarctique*, author's collection.

of whaling boats that it reminded him of a Norwegian fishing port. Although Charcot found the stench of whale flesh overwhelming, when he boarded the largest vessel to meet whaling master Andresen, manager of the Magellan Whaling Company, he was impressed by the comfort and elegance of the ship's interior.

The following day, Charcot and his shipmates embarked on their scientific work. When Charcot noticed a cairn on the hillside, he realised that it might contain a note in from 1905, when *Uruguay* came in search of him after he had failed to return to Buenos Aires as arranged. The bottle in the cairn was broken, but the note was intact. Charcot now read, four years later than Captain Galindez of *Uruguay* had intended:[1]

> Deception Island, 8 January 1905
> This day I have visited this bay with the corvette Uruguay with the object of getting news of the Expedition under the leadership of Dr Charcot, and not having succeeded I am going to Wiencke Island to leave a message there. [signed] Ismael F. Galindez.

Charcot was touched to reach the note and remained grateful to the generous and hospitable Argentinians who had organised coal depots and sent *Uruguay* to search for him. Charcot needed more coal before embarking on his scientific programme, so was grateful when Andresen agreed to supply him with 30 tons immediately, subject to Charcot reimbursing it from his stocks at Punta Arenas. When Charcot asked what he could do to reciprocate, Andresen asked Charcot whether, as there was no doctor at the whaling station, he could examine his wife, who had been feeling unwell, and attend to a man whose hand had been damaged by a cutting machine.

Happily, Wilhelmina or 'Mina' Andresen's ailment was not serious and Charcot enjoyed meeting the first, and probably only, woman he knew of who was living in the Antarctic region and had spent a winter in the south. The injured man's case was more serious, however, and Liouville, Charcot's assistant doctor-cum-zoologist, agreed to amputate some of the man's fingers to avoid gangrene setting in.

On Christmas Day, the whalers' only annual holiday, Charcot invited Andresen, his wife and whaling captains for a farewell champagne reception by way of thanks for their kindness – most recently by agreeing to post mail to France from Punta Arenas and to check cairns at Port Lockroy and elsewhere should Charcot fail to return by January 1910.

'M. et Mme Andresen', Captain Adolphus Andresen, with his wife Wilhelmina ('Mina'), whom Charcot suggested was the first woman to overwinter in Antarctic regions; from p. 317, *Le Pourquoi Pas? dans L'Antarctique* 1908–10, author's collection.

Pourquoi Pas? left Whalers' Bay to a chorus of cheers and ships' whistles and headed south. Charcot made good progress on his scientific programme before overwintering on Petermann Island, reaching considerably further south than he had on his previous expedition. In late November 1909, after the ice went out at Petermann Island, Charcot returned to Deception Island, where he found the Andresens, their parrot and angora cat, and the man with the damaged hand all in good health. After a diver inspected damage to his ship's hull, repairs and a programme of scientific work got under way. There was mail from Punta Arenas, but Charcot's mail bundle had for some reason been missed. He was interested, however, to learn that Ernest Shackleton had set a new Farthest South, his compatriot Louis Blériot had won Lord Northcliffe's prize for the first cross-Channel flight and that Arctic veteran Robert Peary and erstwhile *Belgica* expedition member Frederick Cook were debating publicly as to which of them had reached the North Pole first.

After welcoming 1910 with the Deception Island whalers, Charcot left details of his route at the Pendulum Cove cairn and returned south. Around 70°S, Charcot saw from the crow's nest distant snow-covered peaks rising from an ice cap, which turned out to be uncharted land.

He was just north of Captain Cook's 1774 Farthest South of 71° 10'S, so he wondered whether to continue, but *Pourquoi Pas?* needed repairs and coal so he followed the pack edge to 124°W and then headed for Tierra del Fuego.

In Punta Arenas, Charcot and his men were cheered ashore and after their cables reached France, they were inundated with congratulatory messages. As ever, Charcot felt he could have achieved more, but he had finally surveyed Deception Island and had, thanks to the kindness and cooperation of Andresen and other whalers, obtained sufficient coal to carry out all his work.

After Charcot returned to France, Robert Scott contacted him and suggested Charcot come to England to see his new expedition ship *Terra Nova* and the three motor sledges that Scott was taking to Antarctica.[2] Charcot would have liked to see his friend and exchange news, but as he had expedition work to complete and had been away from home for so long he regretfully declined. Later, however, he mentioned in his expedition report the 'pleasant and profitable time' he and Scott had spent testing motor sledges at Lautaret. He also included few pages on the history of Deception Island and described in some detail his time there and the kindness and cooperation without which his expedition might have been less productive, interesting and enjoyable as it had been.

POLAR POSTSCRIPT: Since Charcot's visit, Deception Island has had a varied and interesting history. A whaling station operated there until the 1930s. In 1928, aviator Hubert Wilkins made the first powered flight over Antarctica. Over the years, Antarctic bases have been established there, including by Britain, Argentina, Chile and Spain. Deception Island suffered several volcanic eruptions between 1967 and 1970, but is now regularly visited during the summer season and is a major breeding area for chinstrap penguins.

32

Matthew Henson's Fur Suit

This fur suit was made for Matthew Henson by an Inuit woman who, with other members of her community, assisted Robert Peary, Henson and other members of the 1908–09 North Pole expedition. Henson first met Robert Peary in 1887 when Peary invited him to join him on a naval survey in Nicaragua, in the capacity of his valet-cum-assistant.[1] Following their return, Peary invited Henson to join him on expeditions to Greenland and the North Pole, during which Henson became, as Peary would describe him, his longest-serving 'assistant'. By 1908, however, Henson was an accomplished dog sledge driver, maintained and repaired sledges and other expedition equipment, hunted and, after learning Inuit, communicated with members of local communities upon whose skills and knowledge he, Peary and other members of their expedition depended.

Matthew Henson's fur suit is in the collection of Berkshire Museum, Pittsfield, USA. The museum was founded by Zenas Crane, Vice President of the Peary Arctic Club.

During Matthew Henson's first Arctic expeditions, he travelled not only with Peary, but with scientists including Frederick Cook and Peary's wife, Josephine, who had given birth to the couple's first child, Marie Ahnighito, at around 77°N. Since then, Cook had travelled to Antarctica and Josephine Peary was raising their daughter and raising funds to cover the costs of her husband's expeditions.[2] Henson had occasionally, when Peary was lecturing or trying to raise funds, accompanied him on tour, but by 1904, Peary had become a leading light within the geographical and exploration establishment.

Matthew Henson's fur suit;
image © and courtesy of Berkshire Museum, Pittsfield, Mass., USA.

In September 1904, Peary, as President of the Eighth International Geographic Congress, welcomed delegates to Washington. During the conference Frederick Cook gave presentations on his time in Antarctica, the differences between the two polar regions and his recent ascent of America's highest peak, Mount McKinley. Cook made no reference to his future plans, but Peary, who was regularly on the stage, mentioned several times that he was planning to return to the Arctic and attempt to reach the North Pole.

In July 1905 Henson, Peary and other expedition members sailed north on SS *Roosevelt*, a brand-new vessel financed by the Peary Arctic Club of wealthy and influential supporters. Initially, all went as planned, but a combination of blizzards and widening leads forced them to return early.[3]

'On the sledge that went to the North Pole', (l to r) MacMillan, Borup, Gushue (1st Mate, *Roosevelt*) and Henson on board *Roosevelt*; image courtesy of Library of Congress.

Peary immediately began planning for his next expedition, but was dismayed when, in August 1907, the Secretary of his Arctic Club received a letter from Cook. The letter, dispatched from Etah Harbour, North Greenland, indicated that Cook was now planning to reach the North Pole by a new route he had discovered during a recent hunting trip with friends.

In early July 1908, SS *Roosevelt* left New York for Etah Harbour, where Peary was told that, while Cook had left some equipment and provisions, there was no further news.[4] In September, *Roosevelt* reached Cape Sheridan, where Henson spent a busy winter making sledges (to Peary's new design), cutting his men's hair, repairing clothing, training sledge dogs, interpreting and, time permitting, reading Dickens's *Bleak House* and Kipling's *Barrack Room Ballads*.

In early 1909, the North Pole party sledged to their final land base at Cape Columbia, where they built igloos and began preparing for the final 400-mile trek to the North Pole. In late February, Peary's men and about 100 dogs set off across the sea ice with Bartlett's group breaking trail and identifying leads.

On 1 April, after reaching 87° 47'N, a new Farthest North, Bartlett and his team turned back, leaving Peary, Henson and four Inuit companions, Ootah, Egingwah, Seeglo and Ooqueah, to continue to the Pole. Peary had spent much of the journey so far at the rear, protected from northerly winds and now he seemed invigorated, even after both he and Henson fell into open leads while crossing sea ice. Peary had never trained Henson to take sightings, but after validating their Farthest North, he seemed happy to trust Henson's judgement on distances until they neared the Pole.

When they pitched camp on 6 April, Peary unpacked a special flag made by his wife, Josephine, which made Henson think this was their final camp. After taking sightings just short of the Pole and a second set beyond it, Peary declared they would plant their flags 'at the North Pole'.[5] After Henson and the four Inuit posed for photographs in their fur suits, cheering and waving flags, Peary cached a strip of his precious flag and a note recording their achievement and they prepared to leave.[6]

To Henson, Peary seemed uncharacteristically subdued on the return journey but they completed it in considerably less time than they had taken on the northern journey.[7] Before *Roosevelt* left Cape Columbia, Peary erected a framed record headed 'Peary Arctic Club North Pole Expedition 1908', on which he listed the names of his travelling companions with the exception of the Inuit, whom Henson regarded as friends.

At Etah Harbour, they learned that Cook had passed through on his way south some time ago, but they could not establish for sure whether he had reached the Pole. When they reached Labrador, however, Peary learned that Cook was claiming to have reached the Pole in April 1908. He became angry and immediately dispatched telegrams to his wife, leading supporters and news agencies insisting that he, not Cook, was the first to have flown the Stars and Stripes at the North Pole. When journalists approached Peary, he downplayed Cook's announcement and insisted that his own claim would, in due course, be backed by proof.

Cook's claim was already under intense scrutiny by British journalist Philip Gibbs, while Peary seemed in no rush to produce his own records. As claim and counterclaim continued, Henson was portrayed (dressed in his fur suit) on a cigarette card and invited as guest of honour to a banquet in Harlem, during which educator and presidential adviser Booker T. Washington presented him with a gold watch.

Henson remained puzzled by Peary's uncharacteristic behaviour during the return journey from the North Pole, which he alluded to in a magazine interview in 1910 and in his own expedition account, *A Negro Explorer at the North Pole*, which was published in 1912. By then, however, Peary's claim was recognised by both America's National Geographic Society and Britain's Royal Geographical Society – although, despite providing Henson with an introduction for his book, he gave Henson relatively little credit in his own published expedition account or during his formal lectures.[8]

POLAR POSTSCRIPT: Henson died, aged 88, in 1955. By then he had, albeit belatedly, been admitted to the Explorers' Club, presented with a Congressional Medal, become the subject of a biography, *Dark Companion*, and had been presented to Presidents Harry Truman and Dwight Eisenhower at the White House.[9] In 1988, he and his wife Anne were reburied in Arlington National Ceremony, Washington DC, where a grave marker shows Henson in his fur suit and describes him as 'Co-Discoverer of the North Pole'. Henson remains an inspirational figure for many, including Dwayne Fields, the first black Briton to walk to the North Magnetic Pole, who nominated Henson for inclusion in the BBC's *Great Lives* radio series.

Part VII

Southward Ho!

In January 1910, as the Cook–Peary North Pole debate continued, Roald Amundsen announced that he was postponing until mid-year his departure from Norway to the Bering Strait, the starting point for his planned drift across the North Pole on Nansen's *Fram*.[1]

In Britain, Robert Scott was preparing for his second Antarctic expedition, during which he planned to carry out a major scientific programme, explore King Edward VII Land and attempt to reach the South Pole. Meanwhile, Scotland's William Bruce and Germany's Wilhelm Filchner were both keen to build on previous oceanic and other surveys of the southern Weddell Sea and, ice permitting, to land and establish whether Antarctica was a single continent, or two or more large land masses divided by frozen straits.

33

Ponting's Kinematograph

In late 1909, photographer Herbert Ponting was offered what promised to be the most exciting and challenging commission of his career – that of photographer and filmmaker on Robert Scott's second Antarctic expedition. Ponting had been a professional photographer since 1901, but he had never operated a kinematograph before, so sought out Arthur Newman, designer of the Newman-Sinclair kinematograph which had been supplied to Borchgrevink's *Southern Cross* expedition.[1] When Ponting left England in September 1910, he took with him both the lightweight metal alloy Newman-Sinclair No. 3 and the sturdier, wooden-cased Prestwich Model 5 kinematographs.

Ponting's Prestwich Model 5 (around 15in by 15in by 7in) is cased in mahogany and weighs approximately 20lb. It is operated by a hand crank and uses 35mm film fed from preloadable film boxes. It is in the collection of the Science Museum Group/National Media Museum, Bradford.[2]

Herbert Ponting's introduction to Scott came through his friend and travelling companion Cecil Meares, whom Scott had already recruited as his expedition's sledge dog trainer and driver.[3] As Scott clearly understood the benefits of professionally produced photographs and film footage (including for raising funds for his expedition), Ponting decided to forgo another commission and join the expedition.[4]

Before Ponting and Scott left Britain, executives at Gaumont's British subsidiary agreed to distribute the expedition films. To ensure the films reached Britain promptly and in good condition, Gaumont would second one of their managers, Fred Gent, to Sydney, where he would take delivery of Ponting's films as soon as *Terra Nova* brought them back from Antarctica each season.[5]

Prestwich Model 5 35mm kinematograph (cine-camera), *c.*1910, (ref. 1928-0984); image © Science Museum Group.

During early 1910, while Ponting juggled existing commitments with procuring photographic equipment for the expedition, Scott agreed that rather than leave on *Terra Nova* on 1 June, Ponting could take a fast steamer from Britain and join them at one of her ports of call.[6] In the event, Ponting just reached Lyttelton, New Zealand, *Terra Nova*'s final pre-Antarctic port of call, in time to set up a kinematograph and film her steaming into harbour. While in Lyttelton, Ponting and Scott finalised arrangements regarding the distribution of press photographs with expedition agent Joseph Kinsey and delivering film footage with Gaumont's Fred Gent, who brought contracts from Sydney for Scott to sign.[7]

As *Terra Nova* left New Zealand, Ponting filmed cheering crowds waving them away. As soon as the ship entered the Southern Ocean, however, he struggled with seasickness as he moved around the heaving

Camera designer Arthur Newman instructing Ponting on using the Newman-Sinclair kinematograph; image (1909–10) from glass negative previously owned by Ponting and/or Newman, © A. Strathie, author's collection.

ship, clutching slippery rails with one hand and a precious kinematograph or camera with the other.

During storms, Ponting's on-deck photographic laboratory was regularly awash, but when *Terra Nova* entered the ice he could film the ship's bow cutting through ice floes from a quasi-gangplank made by ship's carpenter Frankie Davies. When *Terra Nova* was held up by thick pack ice, Ponting clambered down to the ice or up to the top masts to capture views of the ship and surrounding scenery from interesting angles.[8] As for filming wildlife, penguins and seals proved easy subjects, but he soon realised that filming whales' spouts was easier than filming the whales themselves.

When *Terra Nova* entered McMurdo Sound in early January 1911, Ponting worked round the clock, filming and photographing the magnificent mountains, Mounts Erebus and Terror, and the Great Ice Barrier.

Scott decided to build the expedition hut at Cape Evans and he asked Ponting to film men unloading the ship and releasing the ponies and sledge dogs after their long voyage.[9] When filming crates being dragged up the beach became monotonous, Ponting ventured along the cape to examine a huge, strangely shaped beached iceberg. When he looked into the glittering grotto running through it, he realised that the opening perfectly framed *Terra Nova* in the distance.

Ponting's first attempt to film a pod of orcas proved less successful after the sharp-toothed creatures almost butted him and his camera off the ice edge. And while Ponting was grateful for the still long daylight hours, his equipment-laden sledge (which weighed about 200lb) became increasingly difficult to pull through slushy ice and, if the ice gave way, risked dragging him into 1,000ft-deep, orca-infested waters.

Scott's main priority before winter was to lay food depots across the Barrier to support the following season's attempt to reach the pole. As *Terra Nova* would need to leave before Scott returned from leading the depot-laying party, Ponting showed his leader images he thought might appeal to the newspapers and magazines who expected exclusives. Scott had contracted to send Gaumont 2,000ft of film footage (about thirty minutes' running time) per season. As Ponting had already produced 8,000ft and they had no time to go through it, they decided to send it all back with the ship, so Joseph Kinsey could ship it to Fred Gent in Sydney for processing and forwarding to London.

After the depot-laying party and ship had left, Ponting fitted out his darkroom-cum-bedroom, then sledged over to Cape Royds to film the Adélie penguin colony near Shackleton's *Nimrod* expedition hut. As winter approached, Ponting and Edward Wilson, with autochrome colour plates and paints respectively, tried to capture the colours of the pre-winter sunsets.

Ponting, who slept little and preferred to be too busy rather than idle, found the long, cold winter months difficult and, whenever possible, would venture outdoors, whether to help scientists with instrument readings or photograph icebergs illuminated by flash powder. When displays of Aurora Australis began appearing in the sky above Cape Evans, he and meteorologist George Simpson tried to photograph them, but regardless of the shutter speed, exposure or type of glass plate or film, they signally failed to capture images of the huge waves of light which they saw clearly with their own eyes.

There was little scope for filming during the dark months, either outdoors or in the crowded 25ft by 50ft hut, but Ponting developed some earlier films so he could show them after the special Midwinter's Day feast. Although Ponting had never worked as a portrait photographer, he was pleased with his photographs of the men in the hut, whether repairing equipment, cooking or lying in their bunks. Two of his most striking photographs were those of Edward Wilson, Henry 'Birdie' Bowers and Apsley Cherry-Gerrard, one taken when they left for Cape Crozier, the other a month later when they returned, looking gaunt, dishevelled and hungry.

As the sun returned from below the horizon, Ponting resumed full photographic and filming duties. As his heavy equipment effectively precluded him from joining the South Pole journey, he tutored Scott, Bowers and others in taking photographs under the twenty-four-hour Antarctic summer sun. Scott and Ponting were keen that those who saw their films understood 'life on the march', so before the southern party left, Ponting filmed Scott, Wilson, Bowers and Edgar 'Taff' Evans pulling a sledge, erecting a tent and, once inside it, making hot drinks on a Nansen cooker and bedding down for the night.

Scott wanted Ponting to return to New Zealand in early 1912 to deliver films to Gent in Sydney then continue to London to publicise the expedition and help raise funds for another season. In late October, as Scott and the southern party prepared to leave, Ponting accompanied them to Hut Point, so he could film the caravan of ponies, dogs and men processing across the Barrier. Scott knew he and other members of the final South Pole party would probably not return in time to catch the ship. He also knew Amundsen was now trying for the South rather than the North Pole but had decided to continue with scientific and other work as planned.

Ponting wished Scott and Wilson all the best and shook their hands. They wished him well and as they led their ponies away, Ponting began cranking the handle of his kinematograph. Through his viewfinder he saw Wilson pause, turn in Ponting's direction, raise his arm and wave in farewell.

POLAR POSTSCRIPT: When Ponting delivered his second season's films to Fred Gent in Sydney, he learned that Gaumont's executives had been pleased with his first season's films. A few months after Ponting returned to Britain, his second season's films were shown at the London Coliseum and other venues. As Scott and Ponting had predicted, audiences particularly enjoyed the sequences showing Scott and others camping in their tent.

34

A Samurai Sword

Lieutenant Nobu Shirase, leader of Japan's first Antarctic expedition, presented this seventeenth-century Samurai sword to Professor Edgeworth David of Sydney University on 19 November 1911. The sword, given in recognition of David's assistance to Shirase and his men, was made by master swordsmith Hidari Mustu and had been given to Shirase by expedition supporter Tasaburō Fukada 'in admiration of the expedition's courage'.[1]

The sword is in the collection of the Australian Museum, Sydney; it was presented to the museum in 1979 by Edgeworth David's daughter.

In January 1910, Japanese army officer Lieutenant Nobu Shirase petitioned government officials for support for an Antarctic expedition, an initiative which would, he suggested, demonstrate that Japan ranked among the world's 'rich and powerful nation[s]'.[2] Although fundraising proved difficult, Shirase and his men left Tokyo in late November, on *Kainan-maru*, a 100ft-long fishery vessel purchased for them by Shirase's main supporter, ex-prime minister Count Ōkuma.[3]

When Shirase, who knew he was running behind Scott and Amundsen's expeditions, reached Wellington, New Zealand, in early February 1911, he made as quick a turnaround as possible. By early March, Cape Adare was in sight, but when high winds made landing impossible, Shirase continued south.[4] After *Kainan-maru* became trapped in 2ft-thick pack ice at around 74°S, Shirase realised he had no chance of reaching the Great Ice Barrier and depositing a landing party, so he returned north.

Shirase headed for Sydney, where he had already planned to overwinter so he could send reports to Japanese officials along with requests for funds for a second season. Although officials were helpful, press reports

Samurai sword presented by Shirase to Edgeworth David (ref. E76456, Australian Museum); image © and courtesy of Carl Bento (photographer) and Australian Museum.

suggesting Shirase's expedition was a cover for a Japanese surveillance operation made him and his men feel unwelcome.[5] Lacking funds to pay for his men's accommodation while *Kainan-maru* was being repaired, he sought permission to erect his expedition hut in a residential area – and was more than grateful when Professor Edgeworth David of the University of Sydney, who was a member of Shackleton's *Nimrod* expedition, visited him and offered assistance and guidance.

While Shirase and his new mentor planned the forthcoming season's work, ship's captain Nomura took a steamer to Tokyo, delivered Shirase's reports and applied for more funding from Ōkuma and government officials. Shirase had mentioned in his reports that he still hoped to reach the pole but eventually conceded that, rather than compete directly with Scott and Amundsen, he should try for a different Antarctic first – an exploration of King Edward Land. Ōkuma, as before, supported Shirase's efforts and ensured he had additional assistance, including from another scientist, and a kinematographer who would film proceedings including the planned attempt by Shirase's 'Dash Patrol' to land briefly on the Great Ice Barrier and sledge as fast and far as they could to establish Japan's first Farthest South.

In Sydney, while *Kainan-maru* was being prepared for departure, Edgeworth David's protégé, Douglas Mawson (who had accompanied David on Shackleton's *Nimrod* expedition), was now raising funds for his own expedition to Adélie Land and other part-charted areas west

Group on *Kainan-maru* including (front, l to r) Douglas Mawson, Nobu Shirase, Edgeworth David and (back row) unidentified expedition members; image © and courtesy of Australian Museum.

of Cape Adare.[6] Mawson's expedition ship, *Aurora*, was now waiting in Hobart. Before leaving to join her, he visited *Kainan-maru* and members of Shirase's expedition joined Mawson's farewell dinner. In the midst of his own preparations, Shirase wrote to Scott, care of Joseph Kinsey in Christchurch, informing him of his plans and expressing hopes they might, as representatives of allied nations, soon meet.[7]

On 19 November, shortly before *Kainan-maru* sailed, Shirase formally presented David with a precious antique Samurai-era sword by way of thanks for his guidance, support and practical assistance over the past months. Shirase much admired David for his knowledge and his courtesy, but Shirase and his men had also impressed David and won over other Sydney residents with their determination and pluck.

Shirase and his men welcomed 1912 during a Southern Ocean storm, but *Kainan-maru* reached Cape Adare safely and, with cries of 'Banzai!', Shirase's men entered the Ross Sea and continued to the Great Ice Barrier. As they entered the Bay of Whales – where Shirase hoped to land his Dash Patrol and prepare for their sledge journey south – they were engulfed by a blizzard, through which they could just make out a ghostly shape. As they approached, the shape resolved itself into *Fram*, Amundsen's expedition ship. After mooring *Kainan-maru* at a respectful distance, Shirase and his Dash Patrol began unloading equipment. While they did so, Captain Nomura and an English-speaking crew member visited *Fram* to discover

Postcard with Shirase and other expedition members in fur suits, with unknown companion; image © and courtesy of Chet Ross, Chet Ross Rare Books.

that Amundsen's South Pole party were likely, given recent fine weather, to return to the Bay of Whales soon.

On 28 January, the Dash Patrol reached 80° 5'S, where they raised the Japanese standard.[8] When *Kainan-maru* returned to collect them, Captain Nomura informed Shirase that his ship had passed Scott's *Discovery*'s Farthest East, set another Japanese Farthest South and deposited a landing party whose members had become the first men to explore and chart the interior of King Edward Land.

When *Kainan-maru* returned to Wellington, Shirase and his men received a warm welcome. Shirase took a steamer to Tokyo, where he reported to Count Ōkuma and other officials. After *Kainan-maru* arrived, Shirase and his officers were received at the Imperial Palace; following the public release of their expedition films, they also became well-known public figures.[9] Shirase still had financial debts to repay, but he knew his greatest debt was to Edgeworth David, whose support and advice had transformed Shirase's expedition from a potential failure to a national success.

POLAR POSTSCRIPT: Although Shirase's name remains less well known than those of the leaders of other expeditions that overlapped with his, his expedition reports have now been published in English (see Bibliography). There is now a Shirase Memorial Museum in his home town of Nikaho, Akita prefecture, in Japan.

35

Mawson's Anemograph

This anemograph was used to record wind direction during Douglas Mawson's Australasian Antarctic Expedition (1911–14). The anemograph and an anemometer (which records wind speeds) were installed on a ridge near Mawson's main expedition hut on Cape Denison, Adélie Land, where the instruments regularly recorded gales of hurricane force 12 (77mph), beyond the limits of the Beaufort Scale.

The anemograph is in the collection of the South Australian Museum (ref. A60192). The recording arm of the anemograph is inscribed with the name of the makers, Negretti & Zambra of London (No. 266). The reel of paper on the drum is dated 23 February 1912 and marked in orange specialist ink.

In early December 1911, Douglas Mawson and members of the Australasian Antarctic Expedition left Hobart, Tasmania, on *Aurora*, a converted Dundee whaler. As they approached Adélie Land, expedition sledge master Frank Wild spotted a suitable location for their expedition hut, which Mawson named 'Cape Denison' for a major expedition supporter. After the men unloaded everything that Mawson's main landing party needed for the next year, Wild and his smaller 'western party' reboarded *Aurora* so that Captain John K. Davis could deposit them 400 miles further along the coast before returning to Hobart to overwinter.

As Adélie Land was one of the least-explored areas of Antarctica, Mawson had few past records to consult. As winter approached, his anemograph and anemometer readings suggested the gales were coming from the polar plateau, at speeds sometimes approaching 170mph.[1] It was a struggle for meteorologist Cecil Madigan and his helpers to even reach

Mawson's anemograph (ref. A60192); image © and courtesy of South Australian Museum, Australian Polar Collection.

the instruments or take readings, and when they arrived, they often found the stands blown over or damaged by the gales.

Photographer and filmmaker Frank Hurley, always on the lookout for dramatic images, began to follow Madigan and other instrument readers around – while holding firmly to his precious kinematograph and cameras – in the hope of capturing on film scientists battling blizzards and gales.[2] The gales also stymied Mawson's initial attempts to erect a mast and antennae which he hoped would carry the first wireless radio messages sent from Antarctica to an expedition team based at Macquarie Island. If and when received, they would be forwarded to Australia, where funders and journalists, family members and loved ones, including Mawson's sweetheart Pacquita, awaited news.

Photograph of an expedition member leaning on the wind, picking ice (Cape Denison); photograph by Frank Hurley (ref. H138), image courtesy of Mitchell Library, State Library of New South Wales.

As the winter gales eased, sledge parties prepared to leave Cape Denison. Hurley, Bob Bage and magnetician Eric Webb headed south to establish the current location of the Magnetic Pole, leaving Mawson and the others to explore and chart the area between Cape Denison and Cape Adare. As the men left, Mawson emphasised the need to return to Cape Denison by mid-January, when *Aurora* was expected back to collect them.

Of the 'eastern groups', Mawson, Swiss racing skier Xavier Mertz and army officer Belgrave Ninnis faced the round trip. They made good progress until, 300 miles from Cape Denison, Ninnis, their heavy main sledge and six of their best dogs plunged into a deep crevasse. Mawson and Mertz could just see two dogs (one already dead), but there was no sign of Ninnis or the sledge. As most of their ropes were on Ninnis's sledge, all they could do was shout and peer into the crevasse. There was no response, so after Mawson read the burial service, they headed for Cape Dennison with ten days' rations, a small sledge, six rapidly weakening dogs and a one-man tent.

In largely white-out conditions, they ploughed on, day after day. They had no food for the dogs and as their rations ran out, Mawson killed the weakest dog, which he and Mertz (a vegetarian) shared with the remaining dogs. When no dogs remained, Mawson and Mertz began hauling their sledge. As Mertz began suffering stomach pains and other ill effects from his unaccustomed diet, Mawson helped him onto the sledge and began hauling his friend.

In early January 1913, Mertz died.[3] Mawson, now starving and suffering from stomach pains, fell into a 15ft crevasse. He summoned all his strength to haul himself back to the surface and struggled to a cairn, where he found that Hurley and others had deposited rations for him. But before he reached Cape Denison, blizzards swept in and trapped him for another week.

When Mawson reached the hut on 8 February, he learned that Davis had already left on *Aurora* to collect Frank Wild's party, who were on an ice shelf 1,500 miles west of Cape Denison with no permanent hut or rations for another season.[4] As the radio link was now working, Mawson messaged Davis and asked him to turn back and collect him and his six companions. Davis agreed to do so, but when a gale whipped up and prevented Davis from launching auxiliary boats, he radioed again saying he must prioritise Wild's party over those at Cape Denison, who had a hut, a year's rations and a radio link to the outside world.

As Mawson recovered from his ordeal, he learned that Bage, Hurley and Webb had almost reached the South Magnetic Pole and located it at round 84° 43'S. Collectively, they had charted most of the coastline west of Cape Adare and had copious scientific readings, including those of what Mawson felt sure must be some of the most ferocious gales ever recorded.

As darkness settled round Cape Denison and the winds rose again, Mawson and his companions checked their anemograph, anemometer and radio masts regularly. Mawson decided that with time on his hands, he should begin drafting his expedition narrative – but he wondered how he could ever describe Cape Denison's winter gales to those who had not experienced them.

POLAR POSTSCRIPT: Mawson's illustrated expedition narrative and a full-length film based on Hurley's films and photographs were both entitled *The Home of the Blizzard*. One reviewer praised the book as 'a noble record', which described the 'toll of the Antarctic', including 'furious blasts that cover a man's face in ice an inch thick in a minute or two'. The expedition's 'noble record' also included twenty-two volumes of scientific readings and findings, the last of which was published in 1947.

36

Joseph Kinsey's Visitors' Book, April 1912

Joseph Kinsey acted as the New Zealand agent for both Shackleton's *Nimrod* and Scott's *Terra Nova* expeditions. When *Terra Nova* was based in Lyttelton in late 1910, the Kinseys regularly entertained Scott, Wilson, the latter's wife, Oriana, and other expedition members at 'Warrimoo', their home in Christchurch. Kinsey kept a leatherbound visitors' book, to which he added his own or others' photographs of guests. When *Terra Nova* returned to New Zealand in early April 1912, Kinsey looked forward to receiving news of Scott's South Pole journey and entertaining expedition members. April would be a busy month for Kinsey, particularly when he learned of the imminent arrival of an unexpected guest.

The 'Warrimoo' visitors' book is in the J.J. Kinsey Collection, Canterbury Museum, New Zealand (ref. 1940.193.68).

Before dawn on 1 April 1912, *Terra Nova*'s acting captain, Harry Pennell, and expedition secretary Francis Drake landed at Akaroa and delivered Scott's latest reports to the Central News Agency's (CNA) representative, Bertie Hodson.[1] Pennell and Drake knew Hodson had travelled from England specifically to collect the latest reports on what CNA executives hoped would be their greatest scoop for years. All Pennell and Drake could tell Hodson, however, was that ice had forced them to leave McMurdo Sound before Scott's South Pole party returned to Cape Evans. They had Scott's reports, but only up to 4 January, when he, Edward Wilson, Henry 'Birdie' Bowers, Lawrence Oates and Edgar Evans left on the final leg of their South Pole journey.

Hodson also had news for Pennell and Drake: Roald Amundsen had landed in Hobart almost a month ago, claiming to have reached the pole on 14 December 1911. Pennell and Drake took back to *Terra Nova* bundles

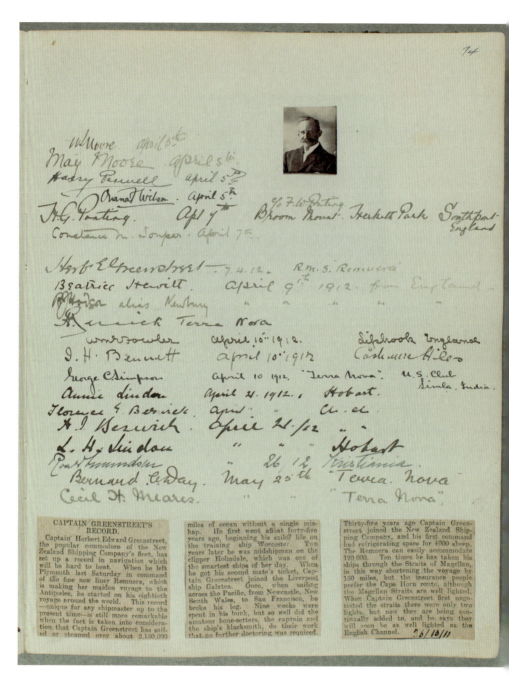

'Warrimoo' visitors' book, p. 74, showing Amundsen's signature (third from foot) and those of others associated with Scott's *Terra Nova* expedition including (from top) Harry Pennell, Oriana Wilson, Herbert Ponting, George Simpson, Henry Rennick, Bernard Day and Cecil Meares; image © and courtesy of Canterbury Museum.

of newspapers with reports of Amundsen's South Pole journey and news of Shirase, Mawson and Wilhelm Filchner's expeditions.[2] By the time *Terra Nova* reached Lyttelton, CNA's news releases had reached newspapers all over the world. Christchurch papers also reported that Scott's naval No. 2, Edward 'Teddy' Evans (whose wife was from Christchurch), was on board *Terra Nova* and might have died from scurvy were it not for the heroic efforts of *Discovery* expedition stalwarts Tom Crean and William Lashly.[3]

Although Amundsen had previously described New Zealand as 'Scott's territory', when he learned that Scott was not back from Antarctica, he agreed to lecture in Auckland, Wellington, Christchurch and Dunedin in late April.[4] Interest in his and Scott's expeditions remained high, but the sinking of White Star Line's brand-new luxury liner *Titanic* with the loss of over 1,500 lives replaced Antarctic news on front pages – so much so that Amundsen was now regularly asked how one iceberg had resulted in the sinking of a supposedly unsinkable ship.

On the morning of Friday, 26 April, Kinsey met Amundsen off the train from Wellington. He took him to meet representatives of Christchurch's Norse community, then brought him to his office, where he introduced him to Teddy Evans. In response to Amundsen's question, Evans confirmed that he had only developed scurvy after leaving Scott and other members of the pole party.[5] After journalists arrived, Amundsen was asked about his long journey and his future plans. Amundsen suggested that, as his postponed drift across the North Pole might last up to five years, he would probably 'take it easy' for a while.

Kinsey then took Amundsen to Lyttelton, where they boarded *Terra Nova* and Kinsey introduced Amundsen to Pennell and his ship's crew. Pennell thought Amundsen seemed quiet and unassuming and concluded that, while the Norwegian's unexpected north-to-south switch had taken Scott by surprise, bygones should now be regarded as bygones.

That evening, Pennell and other *Terra Nova* officers joined Amundsen, Oriana Wilson and other guests at Kinsey's home before attending Amundsen's lecture at the Theatre Royal. The lecture was well received, although the projectionist, somewhat disconcertingly, showed some of Amundsen's hand-coloured lantern slides upside down, sideways or out of order. Back at 'Warrimoo', Kinsey introduced Amundsen to Oriana Wilson, who pointed out that Amundsen's expedition could claim no major scientific discoveries. She seemed pleased, however, when Amundsen assured her that, given her husband and Bowers had survived their winter journey to Cape Crozier, they could probably survive anything.[6]

Photograph of Joseph Kinsey (right) and Amundsen, on p. 76 of 'Warrimoo' visitors' book, which also includes Amundsen's business card with a note of thanks to Kinsey; image © and courtesy of Canterbury Museum

Amundsen's final lecture was in Dunedin and, when he passed through Christchurch on his way back to Wellington, he stopped briefly to make his farewells and present Kinsey with one of his expedition sledges. Kinsey already had a photograph of himself and Amundsen, a business card and Amundsen's signature in his visitors' book but he was glad to have mementos of a visit which, thanks to the efforts of Oriana Wilson and members of Scott's expedition, had gone better than Kinsey expected.[7]

Sledge (wood, leather, zinc) used by Amundsen during depot-laying (ref. 1924.82.1, J.J. Kinsey Collection); image © and copyright of Canterbury Museum.

POLAR POSTSCRIPT: Kinsey's services as New Zealand representative of Scott's and other expeditions were later recognised by a Royal Geographical Society Scott Memorial Medal (1914) and a knighthood (1917). Kinsey's expedition and other papers and memorabilia are held in collections including Canterbury Museum and Te Papa, Wellington.

37

'Three Polar Stars' Photograph, January 1913

On 16 January 1913, members of the Geographical Society of Philadelphia entertained three of the world's most celebrated living explorers to dinner at the Bellevue-Stratford Hotel. All three men were holders of the society's prestigious Elisha Kane Medal: Robert Peary (1902, for reaching the northernmost coast of Greenland), Roald Amundsen (1907, Northwest Passage) and Ernest Shackleton in (1910, following his *Nimrod* expedition). Peary also held a special gold medal, which the society had presented to him in 1909 to commemorate his attainment of the North Pole. During the special event at the Bellevue-Stratford Hotel, the three 'polar stars' posed for photographs, including one that also featured a large globe which had been adjusted in its stand so that Amundsen and Peary could stand beside the appropriate Poles.

The photograph, taken by William H. Rau, shows (left to right) Amundsen, Shackleton and Peary. This copy of the photograph is held by the National Library of Norway, Oslo.

Of the three 'polar stars' who posed for photographs at the Bellevue-Stratford Hotel, Roald Amundsen was uncontestably the man of the moment. He had arrived in New York a week previously and, before lecturing in Philadelphia, had spoken at Carnegie Hall, where Peary had presented him, on behalf of the National Geographic Society, with a special commemorative medal – the southern counterpart to the special medal the society had presented to Peary in 1909.

All three 'polar stars' were much decorated, but all knew that medals did not fund expeditions. Shackleton was currently combining lecturing on his *Nimrod* expedition with promoting his Tabard cigarette business,

'Three Polar Stars' photograph; collection of and image courtesy of National Library of Norway, Oslo (ref. SURA42).

which he hoped might generate sufficient funds to repay his outstanding debts and fund his next expedition.[1] He had been unable to attend Amundsen's Carnegie Hall lecture but had, with good grace, telegrammed him to congratulate him on his 'splendid achievement in the discovery of the South Pole'.[2]

Peary had, completely by chance, met Robert Scott's wife Kathleen in New York on the evening of Amundsen's Carnegie Hall lecture. They had both been beside the elevator in Louis Martin's dining and dancing establishment in New York, where Peary and Amundsen would be attending a supper hosted by the American-Scandinavian Society following the lecture. Peary had previously met Kathleen Scott in London, and she had, in her professional role as a sculptor, designed the Royal Geographical Society's Peary Medal, which Peary had been presented with just before

Scott's expedition ship *Terra Nova* left London. After she explained to Peary that she was at Martin's with friends for an evening's dancing, he had invited her to join him and Amundsen at the post-lecture supper.[3]

Peary understood that Amundsen's decision to turn his attention from the North Pole to the South Pole was partly due to his and Frederick Cook's competing claims to have reached Amundsen's previously stated goal. The gamble had paid off for Amundsen – he had reached his new goal, and his claim was uncontested. For Peary, not only had his North Pole claim been challenged by Cook, but organisations such as Britain's Royal Geographical Society had insisted on copious quantities of paperwork before awarding him a medal. Since the height of the controversy, Cook's claim had been largely discredited, but to Peary's continuing chagrin, the long-established and well-respected American Geographical Society – which had awarded him their Charles Daly Medal in 1902 – had never endorsed or recognised his North Pole claim.[4]

Amundsen's American lecture tour was already proving a great success, but shortly before he had arrived in America newspapers had reported the death, apparently by his own hand, of Hjalmar Johansen, a long-standing friend of Fridtjof Nansen and member of Amundsen's Antarctic expedition team. It had been at Nansen's suggestion that Amundsen had included Johansen as a member of his expedition team, but Amundsen had omitted Johansen from the final South Pole party.[5]

Amundsen had not mentioned the incident publicly or in his expedition narrative, but Nansen was aware of it. Nansen had written an introduction to Amundsen's South Pole expedition narrative and would, Amundsen knew, be distressed at the death of his long-standing friend, particularly in such circumstances. All this added to the sense of moral obligation that Amundsen felt to undertake the North Pole drift he had long since promised Nansen he would make on *Fram*. If Amundsen's lecture tour was successful, he could start making more detailed plans when he returned to Norway later in the year.

Meanwhile, Shackleton knew his *Nimrod* expedition lectures risked becoming outdated soon, particularly after Scott returned to Britain. As the last information on Scott's expedition suggested that a British flag would by now have been flown at the South Pole, Shackleton was already considering another expedition on a grander scale. In the meantime, he awaited news of what Scott had achieved and hoped that his Tabard cigarette venture might generate sufficient profits that he would not need to go cap in hand to sponsors or pledge income from his lectures to discharge loan guarantees.

As the three 'polar stars' went their separate ways following what had been a pleasant interlude in Philadelphia, they wished each other well for the future, whatever that might bring.

POLAR POSTSCRIPT: As those with an interest in astronomy will already know, the celestial equivalent of the three 'polar stars' in the photograph are not static but move and change their designation over time. For example, Polaris, the current northern pole star, became the pole star about 3,000 years ago and will lose that designation during the next millennium. There is currently no single southern pole star, but the Southern Cross galaxy plays a similar role. In theory, both Polaris and the Southern Cross can be seen from the Equator.

38

Henry 'Birdie' Bowers's Sledge Flag

On 18 January 1913, *Terra Nova* returned to Cape Evans, where Teddy Evans, Harry Pennell and others aboard learned that Scott, Edward Wilson, Henry Bowers, Lawrence Oates and Edgar Evans had reached the South Pole on 17 January 1912. As the five had died on the Great Ice Barrier during their return journey, it was now urgent to return to New Zealand as soon as possible to break the news. As men loaded the ship, Evans, Pennell and others carefully sorted and gathered up personal items belonging to the members of the South Pole party. Among Bowers's possessions was a sledge flag that he had asked his family to send from Scotland. When Pennell delivered it to Cape Evans in early 1912, Bowers had already left on the southern journey. Bowers had known his would be the case, but had hoped, after returning to Cape Evans, to fly the sledge flag on the continent he had longed to visit since childhood.[1]

Bowers's sledge flag measures 2ft 6in by 1ft. Made from silk, it incorporates a St George's cross and quasi-heraldic crest.[2] The crest features a half-bent leg pierced by an arrow (possibly a visual pun on Bowers and archers' bows) and the motto '*Esse quam videri*', which means 'to be rather than to seem'.[3] The flag is displayed in the Hall of Memories, Waitaki Boys' High School, Oamaru, New Zealand.

When Henry Bowers was promoted from ship's party to landing party, he was told that he would, based on a long-standing tradition, need a personal flag to fly during sledge journeys. He began making his own flag, but as it looked amateurish compared to the others' sledge flags, he wrote to his mother and sisters in Scotland requesting a 'proper' sledge flag, incorporating a crest his late father had occasionally used.

Bowers's sledge flag in the Hall of Memories, Waitaki Boys' High School; image © Paul Baker, courtesy of Paul Baker and Rector Darryl Paterson, Waitaki Boys' High School, Oamaru.

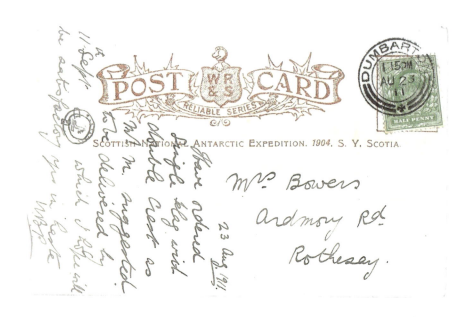

Scotia expedition postcard sent to Mrs Bowers with sketch of crest (reverse shows a Weddell seal); author's collection.

When Harry Pennell brought the new sledge flag south with *Terra Nova* in early 1912, Bowers was already on the southern journey, so he left it with a pile of family letters to await Bowers's return. The sledge flag and letters were still there when Pennell arrived at the Cape Evans hut in January 1913 and learned from his friend Edward Atkinson what had happened since *Terra Nova* left Cape Evans ten months previously.

Attempts to meet up with and relieve the South Pole party before the Antarctic winter had failed and it was not until November 1912 that Atkinson and a search party had found a half-buried tent containing the bodies of Scott, Wilson and Bowers. Scott's journal revealed that Edgar Evans had died in mid-February and that Lawrence Oates, whose feet were by now badly frostbitten and damaged, had walked out of the tent in mid-March rather than delay his companions further. Finally, towards the end of March, Scott, Wilson and Bowers had died in their blizzard-bound tent, apparently from a combination of hypothermia, starvation and dehydration.

Atkinson and the other search party members collected the trio's possessions and, after collapsing the tent over the three frozen bodies, built a cairn over it, which they topped with a cross made of skis. They continued south but found no sign of Oates's body, so built a smaller cairn close to where they found his sleeping bag.

Back at Cape Evans, scientist Frank Debenham processed film rolls that had been found in the tent. One roll had been used, apparently mainly by Bowers, who had photographed Scott, Wilson, Evans and Oates on the plateau pulling their sledge and, later, standing somewhat disconsolately beside a tent Amundsen had left near the South Pole.[4] The other photographs were of all five men, standing and sitting in various formations at the South Pole. In one photograph, Bowers could be seen pulling a thin string with his bare fingers – a duty he apparently shared with Wilson and others as men changed places. The South Pole area looked as bleak and windswept as Scott's journals suggested, but it was possible to make out, beside and behind the five men, a Union Jack, Scott and Wilson's crested sledge flags, Bowers's makeshift flag and a small ensign Teddy Evans had asked Bowers to carry on his behalf to the pole.[5]

On 21 January 1913, *Terra Nova* left Hut Point, where a hilltop wooden cross had been erected in memory of the South Pole party. On 10 February, well before dawn, Tom Crean rowed Pennell and Atkinson ashore at

South Pole photograph showing (l to r) Oates, Bowers, Scott, Wilson, Evans. Sledge flags (l to r): a small flag Bowers carried to the pole for Teddy Evans; Wilson's sledge flag; Union Jack; Scott's sledge flag; Bowers's makeshift sledge flag (Oates and Evans had no flags); from Herbert Ponting's *The Great White South* (opp. p. 280), author's collection.

Oamaru, 150 miles south of Christchurch. After swearing the harbour master and others to secrecy, Pennell and Atkinson sent coded messages to Joseph Kinsey and the Central News Agency, then caught a train to Christchurch and joined Kinsey in his office. By the following day (New Zealand time) Central News Agency had communicated the news to editors of publications all over the world.

Oriana Wilson came to Christchurch immediately, but Kathleen Scott was mid-voyage in the Pacific Ocean and could not be contacted. In America, news agencies contacted Amundsen, Shackleton and Peary for their comments. Amundsen, speaking from Wisconsin, described the deaths as 'Horrible! Horrible!'; Peary, in Washington recovering from surgery, expressed heartfelt condolences to Kathleen Scott and to other relatives of Scott and his companions; Shackleton, now in New York, said he was 'distressed beyond belief' and amazed that Scott, 'the most efficient and most careful of explorers', had died in such a way.

On 14 June 1913, after three years away, *Terra Nova* returned to Cardiff, where Kathleen Scott, her son Peter, Oriana Wilson, Pennell's sisters and a sizeable but respectfully quiet crowd waited on the quayside. Bowers's mother, who had learned of her son's death while on holiday in Rome, did not travel to Cardiff, but in early July she came to London for the unveiling, by Clements Markham, of a memorial plaque to her son on his training ship HMS *Worcester*.

Later in the month, Emily Bowers joined other relatives of the South Pole party and expedition members at Buckingham Palace, where she collected her son's Polar Medal. Also at the Palace was Oriana Wilson, who had advised her on the correct layout for her son's sledge flag and whose husband had travelled to the ends of the earth with her son.[6]

POLAR POSTSCRIPT: In 1939, Bowers's sister May (Lady Mary Maxwell) donated her brother's 'proper' sledge flag to Waitaki Boys' High School in Oamaru, where it was hung, a short distance from the school's memorial to the South Pole party, in the Hall of Memories, which had been built following the Great War.[7] The makeshift sledge flag that Bowers flew at the South Pole is in the Polar Museum Collection (ref. N.273) at the Scott Polar Research Institute in Cambridge.

Part VIII

'White Warfare' and Testing Times

During 1913 and early 1914, Roald Amundsen continued lecturing and fundraising for his postponed drift across the North Pole on *Fram*. While the expedition was in part to fulfil his promise to Fridtjof Nansen, he would, if successful, become the first man to reach both poles.[1]

In Britain, people found the relative proximity of the news of the sinking of *Titanic* and the date of deaths of the South Pole party sobering, particularly given that both had occasioned very similar memorial services at St Paul's Cathedral. Shackleton had, when appearing at the *Titanic* inquiry, spoken of the dangers of sailing through ice zones, but notwithstanding that and the parallels drawn between the deaths of Scott's polar party and Franklin's men, he continued working up his plans for an exploration of the Weddell Sea and trans-Antarctic crossing.

William Bruce had, after struggling to raise funds for his long-planned trans-Antarctic crossing, abandoned the idea in favour of continuing geological surveys and prospecting work in mineral-rich Spitsbergen. Austria's Felix König (who had served on Wilhelm Filchner's Weddell Sea expedition) and Sweden's Otto Nordenskjöld (apparently undaunted by the loss of *Antarctic*) were both considering returning south in the hope of understanding more about the nature and scale of the Antarctic continent.

39

An Expedition Prospectus

On 29 December 1913, Ernest Shackleton confirmed, through the pages of *The Times*, that he would embark on a trans-Antarctic expedition in mid-1914.[1] He had already secured a £1,000 grant from the Royal Geographical Society, but a potential grant of £10,000 from the British government was contingent on Shackleton raising an equivalent amount. As time was short, Shackleton decided against mounting a labour-intensive public fundraising campaign and produced a professionally printed prospectus which he could send to organisations and wealthy individuals with a short covering letter inviting them to request further details or a personal meeting.

The thirty-two-page prospectus is quarto-sized (approximately 9.5in by 12in) and printed in two colours on high-quality paper; this example is from the second print run of the prospectus.[2] Examples of the prospectus can be found in museums, archives, libraries and private collections; a digitised turning-pages version of the prospectus can be found on the National Library of Scotland's website.

Shackleton's prospectus described his trans-Antarctic expedition as the 'natural sequel' to the *Nimrod* and *Terra Nova* expeditions. His main expedition ship, *Endurance* (originally named *Polaris*), became available after an Arctic tourism venture planned by Belgian explorer Adrien de Gerlache and others failed to materialise.[3] His second ship, Douglas Mawson's *Aurora*, would, Shackleton hoped, be captained by John Davis. For land transport, expedition members would use Norwegian skis and sledges, supported by around 100 sledge dogs, and two motorised sledges powered by aircraft propellers.

An Expedition Prospectus

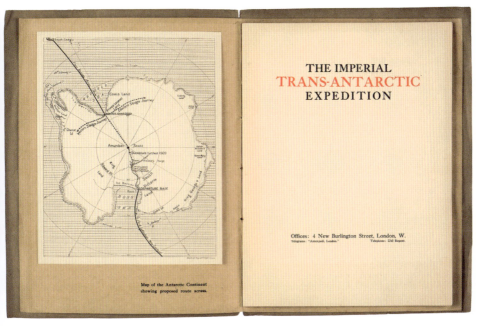

Map showing planned scope and routes of expedition; prospectus frontispiece.

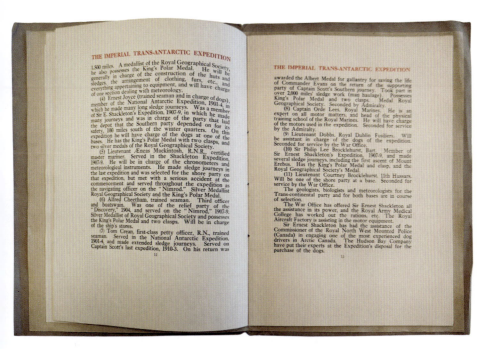

Pages containing information on those who had already committed to join the expedition. Both images © and courtesy of Dominic Winter Auctions, South Cerney, Cirencester.

As to timing, Shackleton planned to establish his main base on the Weddell Sea coast at around 78°S (in practice, Filchner's Vahsel Bay) in late 1914. He and a small party would then, during the Antarctic summer, climb uncharted glaciers to and cross the polar plateau and, via the Beardmore Glacier, join a six-man Ross Sea party, who would already have laid food depots along the Great Ice Barrier and Beardmore Glacier. All being well, Shackleton would complete the crossing by April 1915, but should ice or weather conditions go against them, the crossing would be postponed and take place during the Antarctic summer of 1915–16.

Shackleton had, during the *Discovery* and *Nimrod* expeditions, developed a loyal following, so he could include in his brochure the names of key expedition members. Frank Wild, Shackleton's expedition No. 2, had served on three previous Antarctic expeditions, as had boatswain Alf Cheetham. Seaman Tom Crean, a renowned sledger, had served on two expeditions and others had served on either Shackleton's *Nimrod* or Mawson's *Aurora* expeditions or been seconded from the army and navy. The prospectus named no scientists but made it clear that there would be an extensive scientific programme carried out by those in the three land-based parties and aboard the two ships.

Should any potential sponsor require confirmation of Shackleton's credentials, the final pages of the prospectus contained endorsements and quotations from the presidents of the Royal Geographical Society, Royal Society and their overseas counterparts, a host of 'polar stars', including Nansen, Amundsen, Peary, Nordenskjöld, Bruce and Charcot, and almost forty newspapers and periodicals. In their endorsement, a representative of the *Daily Chronicle* emphasised that 'money given before the expedition starts is infinitely more valuable than any subscribed on its return'. The *Chronicle* had already indicated that, leading by example, it would pay for exclusive publication and film rights. The paper's well-connected news editor, Ernest Perris (whom Shackleton had known for several years), would also assist him by suggesting people to whom he might usefully send his prospectus.

Shackleton and Perris worked together to secure a sizeable 'no strings' donation from Janet Stancomb-Wills (a member of the Wills family of tobacco merchants) and £10,000 towards the purchase of *Endurance* from Midlands-based industrialist Dudley Docker.[4] Smaller donations, including from Scott's friend J.M. Barrie and a member of the Rothschild family, were also promised, but as time ran out, Shackleton pledged future earnings from lecturing and sales of expedition reports to secure bank loan

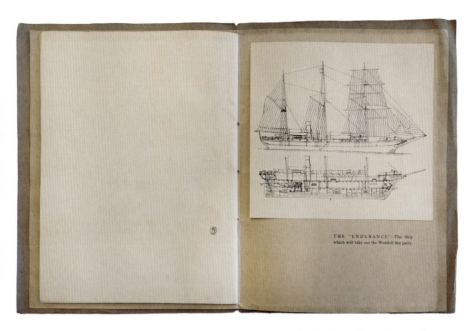

Drawing of *Endurance*, which (as *Polaris*) had been specifically built for polar conditions; image © and courtesy of Dominic Winter Auctions, South Cerney, Cirencester.

guarantees of £5,000 each from Lord Iveagh (of the Guinness family) and Robert Lucas-Tooth (an Australian banker).

In mid-June, Shackleton travelled to Dundee at the invitation of jute industry magnate Sir James Caird, who wanted to know more about the expedition and its finances. Caird initially suggested a donation of £10,000, but after learning Shackleton had pledged future earnings to secure bank loans, he agreed to increase the amount to £24,000 so Shackleton could redeem his two £5,000 pledges and clear other outstanding invoices.[5] Caird agreed that Shackleton could announce details of his donation on the condition that he made it clear that donations from others remained welcome.

Thanks to his prospectus, Shackleton had raised most of the funding he needed. But in late July, as he prepared to leave, Britain's armed forces were preparing for a seemingly inevitable Europe-wide war. After Shackleton released his army and navy secondees, he received the 'all-clear'. On 8 August, *Endurance*, captained by New Zealander Frank Worsley, sailed for Buenos Aires, where Shackleton, Wild and Mawson's photographer and filmmaker Frank Hurley joined the ship.

On 26 October, *Endurance* left Buenos Aires to embark on what Shackleton was now describing as 'white warfare'. As German battleships were reportedly patrolling the South Atlantic, Shackleton bypassed the Falklands and headed for South Georgia, where *Endurance* could be overhauled before entering the Weddell Sea.[6] In another change of plan, Shackleton decided that, rather than risk encountering enemy ships when returning north each season, *Endurance* would overwinter in the Weddell Sea.

Before leaving South Georgia, Shackleton dispatched unpaid invoices to Ernest Perris, a somewhat contrite letter to his wife Emily and specimens of penguins and a letter of thanks to Sir James Caird in Dundee. When *Endurance* left Grytviken in early December, the latest war reports had not arrived but whalers warned Shackleton that the Weddell Sea ice conditions were the worst in years.

Although Shackleton was concerned when *Endurance* encountered pack ice at 57°S, she passed into open water and continued to Coats Land and William Bruce's Farthest South without major difficulties. In mid-January 1915, however, at around 76°S, she became frozen solid into a huge ice floe and began drifting north with the Weddell Sea currents. On 15 February, after failing to free *Endurance* from the pack ice or cross hummocky ice to the Weddell Sea coast, Shackleton bowed to the inevitable and announced that she should now be regarded no longer as an expedition vessel but as 'winter quarters'.

POLAR POSTSCRIPT: In February 1915, *Endurance* was 12,000 miles from London, about 2,000 miles (via the South Pole) from McMurdo Sound, 1,600 miles from South Georgia and beyond the range of wireless radio broadcasts from the Falklands. Shackleton and others who had read accounts of Filchner, Nordenskjöld and Bruce's Weddell Sea expeditions in the ship's library or elsewhere would have been well aware of their precarious situation.[7]

40

A Statue of Cheltenham's Local Hero

A statue of Edward Wilson, who died with Robert Scott and Henry Bowers in March 1912, was unveiled on Thursday, 9 July 1914 in Wilson's home town of Cheltenham, Gloucestershire. He was born, raised and schooled in Cheltenham before moving to London to train as a doctor at St George's Hospital, London. His father, Dr Edward Wilson, was well known in the Cotswolds town as a doctor and in other roles, including as co-founder of its camera club and its museum.

The statue was designed and sculpted by Kathleen Scott, Robert Scott's widow and a well-known sculptor in her own right. It was cast in bronze by monumental sculptors R.L. Boulton & Sons of Cheltenham, who also made the plinth. Following her husband's death and his posthumous knighthood, the sculptor became Lady Kathleen Scott, but she continued to sign this and other works simply 'K. Scott', or (as here) 'K.S.'.

On the afternoon of 9 July 1914, crowds gathered in Cheltenham's Promenade Gardens to watch the unveiling by Sir Clements Markham of a statue of Edward Wilson. During his address, Markham recalled Scott's description of Wilson as possessing the 'keenest intellect' on *Discovery* and Wilson's potential to accomplish 'great things'. Markham also spoke of Wilson's geniality and care for others, his 'heroic resolution' and acceptance of hardship, including during the winter journey to Cape Crozier and return journey from the South Pole.

When thanking Kathleen Scott, Markham praised the statue as a remarkable likeness. While Kathleen was often referred to as 'Scott's wife', she had, long before meeting her future husband at a London dinner party, trained and worked in Paris, including with the great Auguste Rodin.

Edward Wilson's statue, the Promenade, Cheltenham; image © A. Strathie.

Wilson's statue, detail, showing the pouch he used on the South Pole journey and his sledging harness; image © and courtesy of Gill Fargher.

After she and Scott married in 1908, she continued to accept commissions and shortly before he left for Antarctica had designed and sculpted a special Royal Geographical Society gold medal in recognition of Robert Peary's Arctic expeditions.[1] In Scott's absence, she had sculpted busts of Fridtjof Nansen and Prime Minister Asquith and had begun work on a full-length statue of Captain Edward Smith of the *Titanic*.

The statue of Edward Wilson, her husband's best friend, was a labour of love for Kathleen Scott. It was also a collective effort, as Apsley Cherry-Garrard, who had travelled to Cape Crozier with Wilson and Bowers, posed for Kathleen Scott dressed in sledging gear. The 'arms akimbo' stance adopted by 'Cherry' (as expedition members called him) was typical of Wilson and was recorded in numerous photographs by Herbert Ponting. Oriana Wilson and Kathleen Scott had met regularly, including in the latter's studio as the statue took shape, and Oriana had loaned the canvas wallet that Wilson had taken on his southern journey and which had been found on his body with pencils, eraser, drawing paper and other essentials still inside.[2]

The statue's hewn stone plinth was inscribed with a short epitaph. It mentioned Wilson's roles on his last expedition, that he had reached the South Pole in January 1912 and died with Scott just over two months later. The closing words on the epitaph were from the dying Scott's last letter to Oriana Wilson, in which he told her that her husband had 'died as he lived, a brave, true man – the best of comrades and staunchest of friends.'

Like John Franklin's statue in Spilsby, Wilson's statue was erected close to his birthplace, boyhood home and school. It was also close to the museum his father had co-founded and the art gallery where thousands of Cheltonians had recently queued to see his paintings. It was across the road from the W.H. Smith's shop where Wilson had regularly purchased artists' materials and where, more recently, Herbert Ponting's photographs of Wilson and other expedition members had been exhibited in the shop's gallery. It was a few minutes' walk from Cheltenham's Town Hall, where the Wilsons and other Cheltonians had, in 1913, attended a lecture on the expedition by Teddy Evans and, the following year, a cinema lecture by Ponting's friend, Cecil Meares, during which audience members clapped and cheered whenever their local hero appeared on the screen.

Following the unveiling ceremony, the Wilson family, Oriana and guests, including Kathleen Scott, Cherry-Garrard, other Antarctic veterans and Joseph Kinsey and his wife, retired to 'Westal', the Wilson family home, where they could reminisce in private about Wilson and the work

Crowds attending the statue unveiling ceremony; image © The Wilson Family Collection at Cheltenham Borough Council/The Cheltenham Trust.

Photograph taken at 'Westal' following the statue unveiling, with members of Wilson's immediate family and those associated with expeditions, including Joseph Kinsey, Oriana Wilson and Kathleen Scott (all middle row, right side), Clements and Lady Markham (middle row, centre), Reginald Skelton and Louis Bernacchi (back row, left) and Apsley Cherry-Garrard (front row, right); image © The Wilson Family Collection at Cheltenham Borough Council/The Cheltenham Trust.

he had carried out on the continent which had so fascinated him and where he had done some of his most admired and best-known artistic and scientific work.

In late July, Kathleen Scott attended the unveiling of her statue of *Titanic*'s captain, Edward Smith, who had gone down with his ship two weeks after her husband had died on the polar ice shelf. After war broke out in early August, Kathleen combined war work with her sculpture commissions. In 1915, her statues of her husband were unveiled at Portsmouth Dockyards and in Waterloo Place, off London's Pall Mall. She decided that, rather than attend the formal ceremony in London, she would remain at her work bench at the Vickers' munitions factory in Kent, while her son Peter watched the ceremony with friends from the balcony of the nearby Atheneum Club.

Oriana Wilson also moved to the London area, where she worked on a voluntary basis for the New Zealand War Contingent. The two women who had, for a while, been Britain's best-known widows were now, like countless other women, including war widows, simply 'doing their bit for the war effort', as their husbands would have done.

POLAR POSTSCRIPT: Shortly after Kathleen Scott's statue of Wilson was installed, Auguste Rodin left Paris and sought refuge in Britain. He stayed in Cheltenham for several months and, as he walked from his lodgings to the Town Hall to read war news, he regularly passed Edward Wilson's statue. During the latter stages of the war, Kathleen used her sculptural skills to create masks and models for surgeons who reconstructed the damaged faces of war-wounded soldiers, and when the war ended, she sculpted several war memorials.

41

A Rock from Elephant Island

Geologist James 'Jock' Wordie collected this rock and other samples at Elephant Island, where he and another twenty-one members of Ernest Shackleton's *Endurance* expedition lived for months while waiting to be rescued. The small, remote, uninhabited island is part of the South Shetlands group and far from established shipping routes or human habitation. The fact that Wordie and other expedition scientists continued to take readings and collect samples under such conditions is a tribute both to their resilience and dedication to science.

The rock sample is phyllite (metamorphic rock), sheared and cross-cut by quartz veins. It was collected at Point Lookout, Cape Wild on Elephant Island (around 61°S 55°W). It is now in the collection of the Hunterian Museum at Glasgow University (ref. R5083).

When 25-year-old James Wordie joined Shackleton's *Endurance* expedition as geologist he was well prepared. After studying geology at Glasgow University, he had moved to Cambridge, where he attended lectures at St John's College by Raymond Priestley (geologist on the *Nimrod* and *Terra Nova* expeditions) and read scientific reports of William Bruce's *Scotia* expedition to the Weddell Sea. He was an experienced and hardy climber, and was accustomed to collecting rock samples under testing conditions. After being appointed to the expedition, he purchased additional rock hammers and other equipment and, in hopes of being included in Shackleton's trans-Antarctic party, ordered a sledge flag from Kennings & Co., who had made similar flags, including for Priestley's *Terra Nova* shipmate, Henry 'Birdie' Bowers.[1]

By late 1915, Wordie's hopes of collecting rock samples from the Antarctic continent or Weddell Sea coast were beginning to fade. So far,

Phyllite rock sample (ref. R5083. both sides, *c*.6cm by 9cm), collected from Cape Wild, Elephant Island; images © and courtesy of The Hunterian, University of Glasgow.

his geological collection consisted largely of rock samples from South Georgia, small stones retrieved from penguins' stomachs and moraine deposits on icebergs. As the pack ice began to crush *Endurance*, all Wordie could do was follow instructions, gather up his warmest clothing, journal, binoculars, rock hammers and other necessities, and leave the ship. From 'Ocean Camp', their refuge on the pack ice, Wordie and his companions watched as *Endurance* disappeared beneath the ice. Shackleton knew Nordenskjöld and Irízar had replenished food depots on Snow Hill and Paulet Islands but both potential refuges proved frustratingly inaccessible due to currents and ice conditions.

In early January 1916, carpenter McNish and helpers began recaulking and reinforcing *Endurance*'s auxiliary boats, *James Caird*, *Dudley Docker* and *Stancomb Wills*. In early April, following several false starts, Wordie and his shipmates left their last refuge, Patience Camp, and with minimum possessions and maximum possible rations, headed for Elephant Island. Wordie was allocated to the largest and sturdiest vessel, *James Caird*, which was captained by Frank Worsley, but despite Worsley's seamanship it took almost a week of battling through ice floes, pancake ice and huge waves to reach their destination. After feasting on steaks and catching up on sleep, they established a base at 'Cape Wild', where they stocked up on seal and penguin meat and constructed makeshift accommodation.[2]

As there was little chance of ships passing the area, Shackleton announced that he and a small party would sail *James Caird* to South Georgia to seek assistance. On 25 April, Shackleton, together with Worsley, Crean, McNish, Vincent and McCarthy, embarked on their 800-mile journey. Frank Wild would be in charge on Elephant Island until Shackleton returned with a relief vessel and, in the event it became evident that *James Caird* had failed to reach South Georgia, would decide how best to deploy the remaining two boats.

Wordie and others were impressed with the way Wild rose to his new responsibilities, even down to, as Shackleton did, remembering birthdays and other special occasions. On Wordie's twenty-seventh birthday, Wild gave him an afternoon off, during which he climbed, sketched and collected his first rock samples (mainly granite and syenite) since South Georgia and found some 'exotics' on the shingle beaches. Wordie gradually added good-sized specimens from Cape Wild to his collection. Over the months, Wordie concluded that, given the high proportions of phyllite and other metamorphic schists, Elephant Island, geologically speaking, appeared to have more in common with South America than with Weddell Sea coastal areas or Victoria Land.

As winter approached, *Dudley Docker* and *Stancomb Wills* were repurposed as additional shelters. Although temperatures rarely dropped below freezing, stocks of fresh meat ran so low that everyone began looking out for signs of penguins and seals returning. Tobacco was also in short supply and Wordie, who had reduced his smoking to conserve his own dwindling stock, suddenly found himself very popular, to the extent that men showered him with gifts of 'interesting rock samples' in hopes he might offer them a few strands of tobacco or even 'just a puff' of his next smoke.

Wild was determined to keep everyone's hopes alive and had, within weeks of their arrival on Elephant Island, instigated a daily routine of looking out for signs of a relief vessel. He also regularly reminded men to keep expedition records and other necessities to hand in case Shackleton arrived and they had to leave quickly. But it was not until 30 August that a speck appeared on the horizon and began moving towards Elephant Island. After the endless waiting, suddenly it was all a rush, and Wordie

A Rock from Elephant Island 215

Elephant Island, Cape Wild, January 2023; image © and courtesy of Paul Firth.

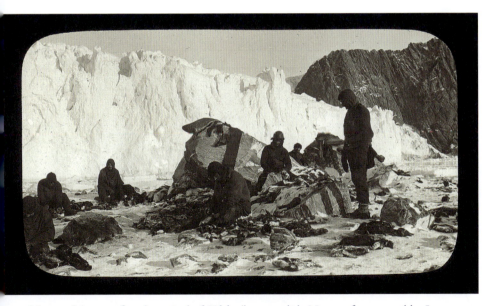

Men packing up after the arrival of *Yelcho* (lantern-slide No. 55 of set owned by James Marr (ref. 1DemXgN9, photograph by Frank Hurley)); image courtesy of Mitchell Library, State Library of New South Wales.

barely had time to gather his notes, rock samples and other necessities from his quarters.

Once aboard *Yelcho*, a Greenock-built Chilean naval vessel, Wordie learned that *James Caird* had reached South Georgia in under three weeks, but *Yelcho* was the fourth ship on which Shackleton had attempted to reach Elephant Island. The war, which had broken out in August 1914, was evidently still raging, which had, in Shackleton's view, made the Admiralty more bureaucratic than ever.[3]

At Punta Arenas, Captain Luis Pardo and his *Yelcho* crew, and Shackleton and his men were welcomed as heroes. Just before Wordie left to return to Britain, Shackleton unexpectedly promoted him to chief scientific officer. In his new role, Wordie was to oversee production of reports by all his fellow scientists, including those in the Ross Sea party, of whom Shackleton had received worrying news, which had preoccupied him for several months.[4]

Back in Britain, Sir James Caird had died in early 1916, without knowing that all the members of Shackleton's Weddell Sea party were safe. In February 1917, while Wordie was preparing for war service, he learned that Aeneas Mackintosh and two other members of the Ross Sea party had died. During the war, Frank Wild's brother Ernest (also of the Ross Sea party) died, as did the ever-cheerful Alf Cheetham. Wordie was injured during the war, but by 1918 was sufficiently fit to return to academic work at Cambridge University and resume work on his geological records, including those from his time on Elephant Island.

POLAR POSTSCRIPT: In 1919, now fully recovered, Wordie joined an expedition to Spitsbergen organised by his fellow Scot, William Bruce, and the Scottish Spitsbergen Syndicate. That year, Frank Wild, James McIlroy and Herbert Ponting were also in Spitsbergen, participating in an expedition organised by another British prospecting company, Northern Exploration Company. In 1920 Spitsbergen, hitherto a *terra nullius* or 'no-man's-land', was ceded to Norway under the Versailles agreement. While mining continues in Spitsbergen today, it never became the 'Arctic Eldorado' that British and other prospecting companies hoped.

Part IX

The Age of Aviation

By the end of the Great War, experienced aviators and newly trained pilots alike had served in army and navy units and in Britain's newly established Royal Air Force.

Tryggve Gran, the *Terra Nova* expedition's ski expert, was already an experienced flyer when he joined Britain's Royal Flying Corps and, under the pseudonym 'Teddy Grant', served in Britain, Europe and northern Russia.[1] His fellow countryman, Roald Amundsen, who embarked on his long-postponed North Pole drift in 1918, had also obtained a pilot's licence before the war and if he became ice-beset, he was prepared to use an aircraft to reach the pole. Shackleton, an aerial pioneer by virtue of his 1902 Antarctic balloon flight, regularly met pilots while serving in north Russia and was impressed by the capabilities of aircraft in Arctic regions.

Australia's Douglas Mawson, who had attempted to use an aircraft in Antarctica in 1911, served in land-based units during the war. His erstwhile expedition photographer Frank Hurley, however, was deployed in Palestine as an aerial photographer and soon became a great flying enthusiast. Somewhat to Hurley's chagrin, however, it was his Western Front brother-in-arms, Hubert Wilkins (who had made pre-war aerial films for Gaumont), who became the first war photographer to receive the Military Cross.

Aviation had come of age during the Great War and many of those who had learned to fly during the war subsequently became full-time aviators. In America, one such was naval officer Richard Byrd, who, after taking part in a US Navy transatlantic aerial crossing, decided his future career lay in the air rather than at sea.

42

An Avro Antarctic Baby

In early 1921, Ernest Shackleton was advised that the Canadian government would not support his proposed expedition to the Beaufort Sea. Within weeks, however, he persuaded the co-sponsor of the now unfeasible expedition, fellow Dulwich College *alumnus* John Quiller Rowett, to finance an Antarctic expedition. Shackleton, keen as ever to deploy the latest technology, contacted ex-Royal Air Force pilot Roddy Carr (whom he had met in north Russia during the war) and ordered an Avro Baby biplane, which was customised for aerial surveying purposes.

Shackleton's Antarctic Baby was the ninth and last Avro Baby manufactured by A.V. Roe of Hamble, near Portsmouth. It was 22ft long, with a wingspan of 26ft and was supplied with floats, a second seat for an observer or photographer and a Marconi three-way wireless system. It had an 80hp air-cooled engine and could cruise at up to 70mph for almost 200 miles.

In September 1921, following frantic preparations, Shackleton left Britain on *Quest* with Frank Wild, Frank Worsley, James McIlroy, Alexander Macklin, Leonard Hussey, Alexander Kerr and other *Endurance* stalwarts. Also aboard were two Boy Scouts (recruited through a national competition), pilot Roddy Carr, Canadian geologist George Douglas and Hubert Wilkins, who would serve as scientist, aerial photographer and navigator.[1]

Before *Quest* sailed, Carr and his helpers disassembled the 'Antarctic Baby' and stowed the fuselage under the bridge. The wings and floats were initially secured on deck but, with space at a premium, they were unloaded at Lisbon so they could be transshipped to Cape Town, where

Avro Antarctic Baby being flown over Southampton Water by Roddy Carr, before he flew it to London to join *Quest*; image © and courtesy of Jan Chojecki.

Shackleton planned to stop en route to South Georgia. When an overhaul of *Quest*'s engines in Rio de Janeiro took longer than expected, however, Shackleton decided to bypass Cape Town and sail direct to South Georgia to begin the pre-winter programme of work. As *Quest* would return to Cape Town to overwinter, Carr remained on board, ready to volunteer for non-flying duties as required.

During the stopover in Rio de Janeiro, expedition doctors McIlroy and Macklin became increasingly concerned about Shackleton's health. McIlroy felt sure that he was suffering from a chronic heart condition, but Shackleton parried questions and declined both advice and medicine. Although he still seemed low on the way to South Georgia, he perked up when New Year's Day 1922 brought calm seas and sunshine.

On 4 January, after they reached Grytviken, Shackleton enjoyed an evening ashore and a game of cards with shipmates before turning in

The Avro stored at St Katharine Docks, London, prior to being dismantled and loaded on *Quest* (in background); image © Topfoto/Central News collection.

early. When Macklin came on watch at 2 a.m. on 5 January, he realised Shackleton was still awake, so went to his cabin. Shackleton, who appeared to be in pain, uncharacteristically asked Macklin for a sleeping draught. Shortly after Macklin returned with the remedy, Shackleton suffered what appeared to be a massive heart attack and died.

The next morning Frank Wild announced, surprisingly calmly, that the expedition would continue as planned, but under his command. While *Quest* was at sea, Hussey would escort Shackleton's embalmed body to Montevideo and await instructions from Emily Shackleton.

As *Quest* got underway, Carr took meteorological readings, taking care to note times and dates when the Antarctic Baby could have flown, whether ahead of *Quest* to identify leads through pack ice or to carry out surveys. At Elephant Island, Wild and other *Endurance* veterans reminisced and regaled Carr and other Antarctic novices with tales of what they had endured while waiting for 'the Boss' to rescue them.

When they returned to Grytviken in April, Hussey greeted them with the news that Emily Shackleton had requested that her husband be buried on South Georgia. The funeral had taken place on 5 March, but before

Bert Hinkler's Avro Baby (the last-known Avro Baby), on display in Bundaberg, Hinkler's home town, courtesy of Queensland Museum, Brisbane; image © and courtesy of Hinkler Hall of Aviation (hinklerhallofaviation.com).

Quest sailed, Wild and fellow *Endurance* veterans climbed a headland overlooking the bay and built a cairn in Shackleton's memory.

As the southern winter approached, *Quest* headed north to the remote island of Tristan da Cunha, where Carr joined Wilkins and George Douglas on scientific surveys. In June, following a brief visit to Gough Island, *Quest* reached Cape Town, where the Antarctic Baby's wings and other transhipped items would be reboarded.

Wild hoped to complete Shackleton's full expedition programme, but while *Quest* was being overhauled in Cape Town, he was informed that due to legal agreements between Shackleton and Rowett, all expedition assets, including *Quest* and the Antarctic Baby, were to be returned to Britain as soon as possible.

As Carr's services were no longer required, he joined Union-Castle's *Walmer Castle*, which reached Britain in late July. When Carr wrote up his meteorological reports, he calculated that even during the expedition's truncated programme, the Antarctic Baby could have flown for around 300 hours, reached 70°S and made aerial and photographic surveys of uncharted land and South Georgia. Although Carr had not tested the

Antarctic Baby in polar conditions, his findings suggested that, as aircraft were now capable of covering the 4,500-mile London–North Pole round trip in a week, the days of polar explorers hauling sledges across the ice or overwintering in hostile conditions were coming to an end.

In mid-September, *Quest* returned to Britain and Carr was notified that he had been accepted for a short-service commission as a Royal Air Force flying officer. As he was supernumerary, he was given leave to enter the Round Britain Race – as the Antarctic Baby was unavailable, Avro provided another aircraft. Carr was delighted to be back in the air after his long absence, even though A.V. Roe's chief test pilot Bert Hinkler outpaced him in another Avro Baby.[2]

POLAR POSTSCRIPT: Following the expedition, Bowring Brothers (who previously acquired Scott's *Terra Nova*) purchased the Antarctic Baby for use in Arctic sealing operations. Although the Antarctic Baby was the last of her class, A.V. Roe later named a new maritime patrol aircraft the Avro Shackleton in honour of the famous British explorer who had hoped to make the first engine-powered flights over Antarctica.

43

A Tribute to Shackleton from a 'Fan'

Dorothy Russell Gregg (later Irving-Bell) claimed to have been 'infected Antarctically' in December 1909, when, aged 13, she heard Ernest Shackleton's *Nimrod* expedition lecture at Bristol's Colston Hall. She continued to attend lectures by Shackleton, Amundsen and other explorers and began collecting related memorabilia. One album page includes two photographs of Shackleton, one from 1909, the other taken in wartime; on that page Russell Gregg wrote lines from Robert Service's 'The Heart of the Sourdough' from which Shackleton often quoted by heart:[1]

> I have flouted the Wild, I have followed its lure
> Fearless, familiar, alone
> But a day must come when the Wild will win
> And I shall be overthrown.

Like many other album pages, this appears to relate to the *Quest* expedition and its aftermath.

The album, now in a private collection, contains autographs, photographs, letters, press cuttings and other material collected between 1909 and the 1930s. Robert Service's words are quoted by kind permission of Anne Longépé, the poet's granddaughter.

Dorothy Russell Gregg, who enjoyed and sometimes wrote poetry, recognised in Ernest Shackleton a fellow poetry lover. He regularly quoted verses, lines or phrases from his favourite poems and poets during his lectures and in his books. Shackleton knew poems by Tennyson and

Shackleton tribute page, showing the year and place of Russell Gregg's first Shackleton lecture.

Dorothy Russell Gregg's invitation to board *Quest* before departure, including permission to bring a camera.
Both photographs by A. Strathie from Dorothy Russell Gregg/Irving-Bell's album, courtesy of Philippa Wordie.

Browning by heart, enjoyed Rudyard Kipling's verses ('If' was a particularly favourite) and was always pleased when someone introduced him to poems by those whose work he had not previously encountered.

A few months before Dorothy attended the lecture in Bristol in 1909, Shackleton had been introduced to the work of Robert Service, the British-born 'Bard of the Yukon' or 'Canadian Kipling', as he was sometimes known.[2] The introduction had come, perhaps surprisingly, in Melbourne, Australia, courtesy of Lady Rachel Dudley, the musically talented wife of Australia's Governor General. She had just received a copy of Service's *Songs of a Sourdough* from a friend in Canada and so much enjoyed them that she read them aloud to Shackleton. Shackleton, equally enthused, tried to commit them to memory and, the following day, recited several to Lady Dudley's friend Ethel Kelly. Before Shackleton left Australia, he wove a quotation from Service into a letter to his friend, Hugh Mill.[3]

Back in Britain, Shackleton began using phrases like 'the Great White Silence' or 'the Great Alone', phrases Service used to describe the Yukon, but which seemed equally applicable to Antarctica. Shackleton, a natural wordsmith, also had no compunction, as and when required, in substituting 'south' for Service's 'north' to fit the context and continent.

Members of the recently formed Poetry Society (which promoted the work of living poets) duly noted Shackleton's enthusiasm for works by Service and other contemporary British-born poets. After he agreed to become their honorary vice president, he was invited to address their members. In his talk 'Poetry in Active Life', Shackleton praised Service and other poets who wrote of 'the outdoors' and described *Discovery* and *Nimrod* expedition members reading and reciting poetry and, in his role as editor of the *Discovery* expedition's 'South Polar Times', finding men's own recently composed poems in his 'Editor's Box'.

Dorothy Russell Gregg had, over the years, sent postal orders to Shackleton and other explorers after they appealed for funds to support their expeditions. In 1921, shortly before *Quest* sailed, she travelled to London's St Katharine Docks, where she was allowed to board the ship and photograph Shackleton and other expedition members. In 1922, she added to her collection a copy of the order of the St Paul's Cathedral memorial service for Shackleton and travelled to Portsmouth to welcome Frank Wild and others when they returned on the *Quest*.[4]

Dorothy (or 'Squibbs' as she was nicknamed) mourned the loss of Shackleton, but she was still was also a 'party girl' and on 7 February 1923 she joined Frank Wild, Frank Worsley, James 'Scout' Marr and other

(Left) Programme for Chelsea Arts Club Ball.
(Right) Dorothy 'Squibbs' Russell Gregg and James 'Scout' Marr (both nicknames were widely used) at Chelsea Arts Club Ball; both photographs by A. Strathie from Dorothy Russell Gregg/Irving-Bell's album, courtesy of Philippa Wordie.

Shackleton expedition veterans at the Chelsea Arts Club Ball. As one of the themes was 'Antarctic', Wild had provided a sledge, real sledge dogs and an image of *Quest* for projection on the main stage.

During the ball, which lasted from 10 p.m. until 5 a.m., *Quest* expedition sponsor John Rowett and his wife entertained a table of guests and Wild and his wife wore penguin costumes. James 'Scout' Marr wore a head-to-foot bearskin, which seemed less appropriate than Dorothy's much-admired silk iceberg costume, but everyone enjoyed dancing the 'Penguin Walk' to a new, specially commissioned tune.[5]

Hugh Mill's *The Life of Ernest Shackleton* was published the same year. Mill regularly referred to Shackleton's love of poetry and each chapter was headed by a few lines of verse. Browning's work (to which Emily Shackleton had introduced her late husband) featured most, but the chapter entitled 'Shackleton Attains, 1908–9' was headed by an extract from Service's 'The Lure of Little Voices', which referred to the 'vast and God-like spaces/The stark and sullen solitudes that sentinel the Pole'.[6] In the subsequent chapter, Mill recalled Emily Shackleton telling him that when Shackleton quoted Service's phrase 'the trails of the world be countless' during a banquet in Norway in late 1909, she had looked at Amundsen and seen the expression on his face change. As she suggested to Mill, she had probably witnessed the moment Amundsen resolved to try to reach the South Pole rather than wait for the North Pole controversy to be resolved.[7] Mill also suggested that Shackleton, probably more than any explorer, had understood the power of words. The late explorer had apparently described the living poet as 'the one man in the world who brings home the glamour and the mystery of the unknown'.[8]

Dorothy Russell Gregg, who shared Shackleton's love of poetry, had, in her own choice of Service quotation for her album, paid tribute both to Shackleton's love of 'the Wild' and his admiration of the poet through whose words Shackleton could express to others what 'the Wild' meant to him.

POLAR POSTSCRIPT: In 1924, Herbert Ponting's full-length version of his *Terra Nova* films was released under the Service-inspired title *The Great White Silence*. By then, members of Shackleton's expeditions regularly included lines from Service's poems in book dedications for many years after Shackleton's death. In 1926 Dorothy Russell Gregg published her own poetry collection, *Chaff*, which was described by a reviewer as 'pleasing, unpretentious verse'.

44

'Uranienborg': An Explorer's Refuge

On 12 May 1926, after Roald Amundsen, Lincoln Ellsworth, Umberto Nobile and their companions overflew the North Pole on airship *Norge*, Amundsen became the first man to reach both poles, as well as traverse both the Northwest and Northeast Passages. In an age of mass communication, Amundsen was world famous, but as was his custom following expeditions, he returned to 'Uranienborg', the fjordside home he had purchased in 1908, where he spent time with family and friends and, thanks to his collection of pictures and memorabilia, could reflect on almost three decades of polar exploration.[1]

'Uranienborg' is in Svartskog, south of Oslo; the name refers both to the Uranienborgveien district where Amundsen was raised and the observatory and home of the celebrated Danish astronomer, Tycho Brahe. In addition to living accommodation, it included outbuildings where Amundsen could store expedition equipment. 'Uranienborg' is now a museum; it first opened in the 1930s and is managed by MiA/Museums in Akershus, who operate seasonal openings and a dedicated website.

On 16 July 1926, at 'Uranienborg', Roald Amundsen entertained family members and local friends for what became a fairly riotous celebration of his fifty-fourth birthday and his recent flight over the North Pole.[2]

Following the pleasant interlude, Amundsen completed his *Norge* expedition narrative and, after collecting reports from Ellsworth and other expedition members, submitted the texts to his publishers. That done, he began drafting an autobiographical memoir that his American publishers, Doubleday & Co., had commissioned during his visit to America the previous year.

'Uranienborg' from the fjord (Anders Beer Wilse, ref. 9982); image courtesy National Library of Norway.

The living room, showing 'A Very Gallant Gentleman' (left) and Millais's 'North-West Passage' (right) near doorway (ref. NPRA113); image courtesy of National Library of Norway.

At 'Uranienborg' Amundsen had everything he needed to reflect on and write about his life: a large, well-appointed study, expedition reports and records and, scattered throughout the house, mementos of his expeditions. A stuffed Adélie penguin reminded him of his first polar expedition and his friend Frederick Cook, who had taught Amundsen the importance of eating fresh meat during polar winters. The invoice from his purchase of *Gjøa*, his first expedition ship, recalled his long traverse of the Northwest Passage and his learning, in a telegraph station in the Yukon, that Norway had been granted full independence. A model of a Dornier Wal flying boat suspended from his living room ceiling was a memento of his first attempts to reach the North Pole by air.

Of the paintings lining the walls of Amundsen's house, he was fond of an oil painting of *Fram* that had been presented to him in 1913 by American artist Frank Stokes, who was now a friend and had joined Amundsen and Ellsworth on *Norge*.[3] But two other pictures, both reproductions, held even more significance for Amundsen and hung on his living room wall, where he would see them every day when at home. The older of the two works, 'The North-West Passage' by John Millais, recalled Amundsen's long-standing fascination with Arctic exploration, which had its roots in his reading accounts of John Franklin's expeditions at the age of 15. The other picture, a reproduction of Charles Dollman's 'A Very Gallant Gentleman', showed Captain Lawrence Oates of Scott's South Pole party limping into a blizzard in the hope that, by sacrificing his own life, Scott, Edward Wilson and Henry Bowers might survive. It was, as Amundsen admitted to his favourite nephew Gustav, one of his most treasured possessions.[4]

Amundsen did not, despite best intentions, complete his memoir over the summer, and in November left 'Uranienborg' for New York, where he accepted a lecturing engagement. On arriving, Amundsen learned that *Norge*'s designer and pilot, Umberto Nobile, was giving lectures which described the *Norge* expedition as a triumph for himself, Italy and Benito Mussolini. This was not the first time Nobile had, Amundsen felt, claimed more than his fair share of credit for the expedition.

Rather than ignite another North Pole controversy, Amundsen postponed his lectures and accepted his publisher's offer of a 'writer's den' at the Waldorf Astoria, where he could complete his memoir. With no distractions, Amundsen completed his draft by Christmas – but warned his lawyer that some passages might be considered 'nasty'.

After Nobile returned to Italy in early 1927, Amundsen gave several lectures and promoted his *Norge* expedition account in America.

Amundsen's reproduction of Charles Dorman's 'A Very Gallant Gentleman'; image © and courtesy of Roald Amundsen's House, MiA/Museums in Akershus.

Amundsen boarding the Latham aircraft; image courtesy of National Library of Norway.

He then accepted an invitation to visit Japan, where he was received at the Imperial Palace, showered with honours and thoughtful gifts (including a handmade *yukata* robe with polar bear motifs) and met Nobu Shirase, who had, like Amundsen, moored his ship in the Bay of Whales fifteen years previously.[5]

Amundsen returned to 'Uranienborg' in August. He had warned his lawyer about 'nasty' passages, but when his memoir *My Life as an Explorer* appeared, headlines and reviews focussed less on his achievements than on his criticism of Umberto Nobile and other explorers. 'Amundsen angry', suggested one reviewer. Amundsen also criticised organisations including America's National Geographic Society, whose officials had cancelled (unjustifiably, in Amundsen's view) his lecture to their members following his prison visit to his old friend Frederick Cook.[6] But the 'nasty' remark which generated the most headlines was Amundsen's suggestion that the British were 'very bad losers' – a judgement that was based on his recollection of comments made by the Royal Geographical Society's late president, Lord Curzon, at a dinner in Amundsen's honour in 1912.[7]

My Life as an Explorer, perhaps inevitably, was not the success Amundsen and his publishers hoped for when they had discussed the commission. Amundsen's spirits dipped further when, in spring 1928, Nobile announced he was planning a North Pole expedition on his latest airship, *Italia*.[8] On 24 May, the day Nobile left Spitsbergen, Amundsen welcomed Hurbert Wilkins and other guests to 'Uranienborg' for a grand celebration of Wilkins's recent Alaska–Spitsbergen flight. A few days later, Amundsen was informed that wireless radio contact with *Italia* had been lost.

Putting aside his grievances, Amundsen confirmed that he was willing to join the search effort. While he waited for an aircraft to become available, Tryggve Gran and other Norwegian explorers, a team on Amundsen's erstwhile baseship *Hobby*, a team on *Quest* and Jean-Baptiste Charcot on *Pourquoi Pas?* also joined or enlisted for the search. By early June, hopes were fading for Nobile and his men, but shortly after Amundsen was advised that a Latham seaplane was available for his use, a faint radio signal suggested that there was a chance they were alive.

At 4 p.m. on Monday, 18 June, Amundsen and his crew took off from Tromsø on Latham 47.02. A few hours later, from somewhere between Tromsø and Spitsbergen, there was a radio signal from the Latham. There were no further messages from Amundsen, but such was his reputation and experience that all attention remained focussed on the search for *Italia* survivors. As they were finally located, concern grew for Amundsen and others in the Latham.

American heiress Louise Arner Boyd, who had chartered *Hobby* for an Arctic summer cruise, agreed that if experienced searchers joined her and her friends, she could now search for Amundsen. But despite *Hobby* covering some 10,000 miles and Charcot and others' best efforts, there was no sign of Amundsen, the Latham or her French crew.

In mid-July, a reporter from Oslo's *Dagbladet* interviewed Amundsen's nephew, Gustav, at 'Uranienborg'. While the pair were in Amundsen's living room, the journalist asked Gustav about the picture of the man in the blizzard. Gustav told the reporter about Oates's heroic attempt to save the lives of his companions and explained that it was among his uncle's most treasured possessions.

That day, Gustav told another reporter not to assume his uncle was dead. Six weeks later, after wreckage from the Latham was found, Gustav and other members of Amundsen's family accepted that the seemingly indestructible explorer would never return to 'Uranienborg'.

POLAR POSTSCRIPT: In October, at a memorial ceremony in Oslo, Fridtjof Nansen spoke movingly and emotionally of his erstwhile protégé. Among those present was Louise Arner Boyd, whose efforts in searching for Amundsen resulted in her being awarded Norway's Order of St Olaf and France's *Legion d'honneur*. What she had planned as an Arctic pleasure cruise became the first of a series of Arctic and other scientific expeditions that culminated in her becoming the first female board member of the American Geographical Society.

45

Mawson's Gipsy Moth

In 1929 Douglas Mawson took a de Havilland Gipsy Moth floatplane and two Royal Australian Air Force pilots, Stuart Campbell and Eric Douglas, on his British Australian and New Zealand Antarctic Research Expedition (BANZARE). Although Mawson's fellow Australian, Hubert Wilkins, had made the first flight over Antarctica in 1928, BANZARE would still be a historic expedition, as the British government had loaned Mawson Scott's *Discovery* and seconded James Marr (*Quest* expedition's 'Shackleton Scout'), who was now an experienced polar scientist. Mawson had also secured the services of Frank Hurley, now one of the world's most famous travel photographers and filmmakers, and naval officer William R. Colbeck, whose father, as captain of *Morning*, had relieved *Discovery* in 1904.

The images of the Gipsy Moth were taken by Frank Hurley and made into lantern slides for pilot Eric Douglas, who regularly took Hurley up in the Moth. They are held, along with Douglas's other expedition photographs, his Zeiss Ikon Icarette camera, pilot's jacket and other material, in the Eric Douglas Antarctic Collection at Museums Victoria, Melbourne.

When *Discovery* reached Cape Town in October 1929, the Gipsy Moth was still in its component parts. After pilots Stuart Campbell and Eric Douglas joined the ship, they and a team of helpers assembled the aircraft, which was then stowed on pallets above the deck.[1]

Around the Antarctic coast, weather, ice and sea conditions prevented any flying until 29 December, when, close to the Antarctic Circle, the Moth was lowered from *Discovery* and prepared for take-off. During their inaugural flight, Mawson and Campbell soared to 5,000ft, from where Mawson saw

The Gipsy Moth stowed on *Discovery*'s deck.

The Moth swinging from the hoist with Douglas (l) and Campbell (r) on the wings. Images from the Eric Douglas Antarctic collection (includes lantern slides, negatives and photographs by Frank Hurley and Eric Douglas), Antarctica Collection, Museums Victoria; images courtesy of Museums Victoria, Melbourne.

uncharted land to the south and what appeared to be leads in the pack ice. While it was an encouraging start, the fine weather was followed by a hurricane, during which the Moth's wings were damaged.

On 13 January 1930, while the Moth was being repaired, Mawson and a small group made the expedition's first landing on an outcrop off Enderby Land. In line with his instructions from British and Australian government officials, Mawson raised the Union Jack, read out a prescribed proclamation and deposited a notice to the effect that the land south of 65°S and between 47° and 73°E was now British sovereign territory.[2]

After being alerted to the presence of a Norwegian ship in the area, Mawson made radio contact with her captain, Amundsen's friend Hjalmar Riiser-Larsen. Following an amicable meeting aboard *Discovery*, they agreed that the Norwegians would operate west of 45°E, while Mawson remained east of the invisible demarcation line.[3]

As the Moth was now back in service, Campbell and Mawson made an aerial survey of Proclamation Island, then handed over to Hurley and Douglas so they could make a flight. Hurley, now in his mid-forties, was still irrepressible and, after filming a huge ice plateau, encouraged Douglas to fly over to a range of mountains which soared 2,000ft above the Moth's altitude of 5,000ft. Although the fine flying weather seemed set to continue, *Discovery*'s coal stocks were running low, so they returned to the Kerguelen Islands, where they had previously deposited sufficient coal for the return voyage to Hobart.

While *Discovery* was overhauled for the 1930–31 season's voyage south, Douglas and Campbell returned to regular air force duties. In Sydney, Hurley edited aerial and other footage into a short film, *Southward Ho! with Mawson*, which played in local cinemas with a live commentary by Hurley or hydrologist Alf Howard. As audience reaction suggested to Hurley that he needed more light-hearted footage for the full-length expedition film, when *Discovery* stopped at Macquarie Island in late 1930 he filmed penguins' antics and (against Mawson's advice) Douglas and Campbell astride huge elephant seals.

As *Discovery* approached Cape Denison in early January 1931, the gales reminded Mawson and Hurley of their last expedition together. Thankfully, wind speeds dropped sufficiently to make a landing, during which Mawson duly made another proclamation and took magnetic and other readings to compare with his *Aurora* expedition records.[4] Although Mawson considered both science and proclamations a serious business, Hurley spent much of the landing filming penguins crowding round a

View from the Moth over pack ice to Proclamation Island and the Antarctic mainland; image from lantern slides/glass negatives (by Frank Hurley, Eric Douglas) in Eric Douglas Antarctic Collection, Antarctica Collection, Museums Victoria; image courtesy of Museums Victoria, Melbourne.

model of Mickey Mouse and a portable His Master's Voice record player he had brought onshore in the hope they would react to the music.

By mid-January, as gales died down, the Moth was back in action. Campbell and cartographer Karl Oom reached 8,000ft and saw land about 100 miles to the south, but rather than waste precious coal trying to reach the land by ship, they continued the aerial surveys. The following week, it was too cloudy to fly, but on 27 January, Mawson and Douglas went up and flew over Wilkes Land. The flight ended badly, however, when a gust of wind smashed the Moth into the side of *Discovery* as it was being hoisted aboard. Although Mawson, who was standing on the hoist, managed to avoid being pitched into the freezing water below, the Moth was badly damaged. Hurley, however, who had been filming the landing and aftermath from the crow's nest, was delighted to have been on hand to capture footage he knew would thrill cinema audiences.

Douglas, Campbell and helpers soon had the Moth back in the air (albeit now with a starboard list) and, although there was little to see but ice, basking sea leopards and the occasional whaling ship, Campbell and

Oom returned from one flight having charted land stretching to 69°S. At their next port of call, MacRobertson Land (named for the expedition's main commercial sponsor), Mawson made another declaration of sovereignty.

Discovery's coal stocks were again dwindling, so Mawson reluctantly abandoned plans for a full aerial survey of Enderby Land and drew the expedition's second season to a close. *Discovery* returned to Hobart after covering almost 11,000 miles in four months. Mawson was satisfied with the results of aerial surveys, given ice and weather conditions, while Hurley had enjoyed his filming flights with Douglas. For his part, Douglas – a keen amateur photographer – was delighted to have flown with one of the world's great travel photographers, Frank Hurley, and to have avoided ditching the Gipsy Moth into freezing Antarctic waters.

POLAR POSTSCRIPT: In late 1935, when Amundsen's *Norge* expedition partner Lincoln Ellsworth and others were reported missing during a trans-Antarctic flight, Eric Douglas was asked (at Mawson's suggestion) to lead a rescue attempt. Thanks to Douglas, his team and another de Havilland Gipsy Moth, the Americans were spotted and rescued from the Bay of Whales, where American aviator Richard Byrd had established a series of 'Little America' land stations.

46

A Young Explorer's Special Medal

In early summer 1932, Henry George 'Gino' Watkins, leader of the 1930–31 British Arctic Air Route Expedition, became the youngest person to receive the Royal Geographical Society's prestigious Founder's Medal. Watkins was then an undergraduate at Cambridge University, where he was mentored by polar scientists Frank Debenham, Raymond Priestley and James Wordie. All three saw in Watkins and his friends a new generation of explorers who might achieve great things in the polar regions.

Gino Watkins's full set of medals includes the Founder's Medal, Royal Danish Geographical Society's Han Egede Medal, the Royal Scottish Geographical Society's Bruce Medal and the Polar Medal.[1] All four medals, currently in a private collection, were included in an exhibition organised by Spink & Son, London, in 2019; they also featured in an accompanying publication (see under Roan Hackney in Bibliography).

By 1931 Henry 'Gino' Watkins was regarded as the standard bearer for a new generation of British polar explorers. Still in his early twenties, he had, before leading the British Arctic Air Route Expedition, climbed extensively in continental Europe, learned to ski, obtained his pilot's licence, organised and led ground-breaking expeditions to Svalbard and Labrador and been elected a Fellow of the Royal Geographical Society. But when not writing expedition reports or lecturing to Royal Geographical Society members, Gino led the life of a Bright Young Thing – dancing the night away with friends in London clubs, scaling the roofs and towers of Cambridge colleges, motoring around the countryside and enjoying house parties, including at his uncle's Cotswolds mansion at Dumbleton.[2]

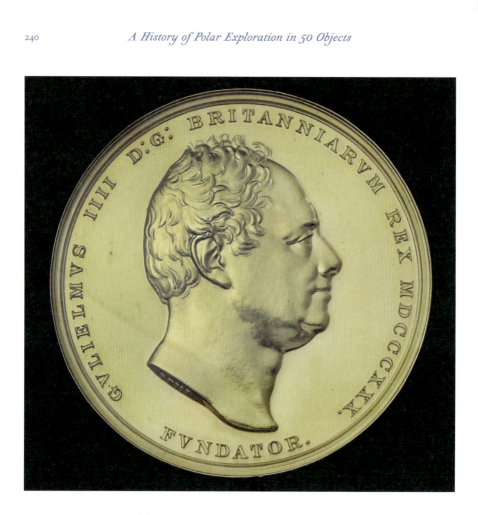

Gino Watkins's RGS Founder's medal.

Gino Watkins's Polar medal (obverse and reverse).

Images © Spink & Son, courtesy of Spink and the owner of the medals.

The main aim of the Royal Geographical Society-sponsored British Arctic Air Route Expedition was to assess the feasibility of a potential air route from Britain to Canada's west coast. As the shortest route involved crossing the largely uncharted Greenland ice cap, Watkins and his fourteen-man team of Cambridge-based friends, volunteers and army, navy and Royal Air Force secondees had established a year-round meteorological station on the ice-cap summit.

While one or two men at a time monitored the ice-cap station, two sledge parties crossed and charted the interior of Greenland, and others surveyed over 1,000 miles of coastline from expedition ship *Quest*, two aircraft or small motor boats and kayaks. Unfortunately, the loss of one aircraft resulted in the cancellation of planned test flights from Greenland to Canada.

Although Augustine Courtauld only just survived his heroic solo overwintering at the polar cap weather station, Watkins and his men sustained themselves by hunting, including in kayaks, using techniques that members of Inuit communities based near the expedition's base camp at Angmagssalik had shown them. But while Watkins and his companions learned much about survival techniques, the news of the death of polar scientist Professor Alfred Wegener and his companion on the ice cap was a sobering reminder of the perils of polar exploration.[3]

Back in Britain, Watkins attended welcome home parties, was presented to the king, lectured to Royal Geographical Society members and enjoyed a family Christmas at Dumbleton. In London and Cambridge, he wrote up expedition reports and planned his next expedition.

Following discussions with Debenham and Wordie, Watkins shelved plans for a fourth Arctic expedition in favour of a trans-Antarctic crossing. On 9 January 1932, he accompanied Debenham to the unveiling of a statue of Shackleton at the Royal Geographical Society and to the fourth annual dinner of the Antarctic Club.

Watkins's initial plans included not only the trans-Antarctic crossing but an aerial survey of uncharted areas of the south-west Weddell Sea coast. The trans-Antarctic party would bypass the pole, thus maximising the amount of new ground covered. Reactions from members of the Antarctic Club and others were favourable, but Watkins was still in debt from his last expedition and now had to raise around £40,000.

Watkins began spending most of his time lecturing or writing to potential expedition sponsors. Even those who appeared interested in his plans would cite the continuing depression in their polite refusals and while the *Discovery* Committee offered to provide Scott's erstwhile vessel for a

Watkins hunting in his kayak with a harpoon rifle; from opp. p. 300 in J.M. Scott's *Gino Watkins*, author's collection.

nominal fee, they could offer nothing towards running costs. Eventually, Watkins settled for the considerably cheaper *Quest* and scaled down his trans-Antarctic party from eight men to four – and took some comfort from the fact that even America's most acclaimed Antarctic aviator and explorer, Richard Byrd, was struggling to raise funds for his next expedition.

While still fundraising, Watkins was invited to Copenhagen, where he was presented with the Royal Danish Geographical Society's prestigious Hans Egede Medal – of which he and Roald Amundsen were among the few non-Danish recipients. Back in London, while his funding application to the Commonwealth Office remained 'pending', he was grateful for offers of modest donations and loans of equipment from companies. Although the *Discovery* Committee were now prepared to loan *Discovery* free of charge, with no funds to cover the running costs Watkins had to decline the offer.

In May, Watkins learned that he had been awarded the Royal Geographical Society's Founder's Medal and would be the youngest person to receive the honour. While that and attendances for *Northern Lights* (the film of his Greenland expedition) were gratifying, Debenham's letter

(L to R) Arctic explorer Vilhjalmur Stefansson, Watkins and Admiral Sir William Goodenough (then RGS President) on Watkins's expedition ship *Quest*; from opp. p. 190 in J. M. Scott's *Gino Watkins*, author's collection.

to *The Times* about the expedition generated only small donations rather than the £10,000 needed. As Watkins's window for reaching Antarctica in time to make the most of the 1932–33 summer closed, he considered accepting an offer from Pan-American Airways for additional paid work on the trans-Greenland air route.

In June, Watkins was awarded the Royal Scottish Geographical Society's Bruce Medal. His mentor James Wordie, the inaugural recipient of the medal, helped Watkins produce a plan for a scaled-down expedition to Graham Land, but despite the *Discovery* Committee's offer of £3–4,000, he still needed another £10,000. Watkins decided he had no option but to postpone his Antarctic plans and return to the Arctic. He accepted Pan-American Airways' offer of £500, which he supplemented with a £200 Royal Geographical Society grant, £100 from *The Times*, loans of instruments and a publisher's promise of £500 for a book on his somewhat unexpected Arctic expedition.

After Watkins had been formally presented with the Royal Geographical Society Founder's Medal, he left Britain with his fiancée, aviatrix Margaret Graham, and his friends John Rymill, Quintin Riley and Freddie Chapman. On 14 July, after a short but enjoyable stay in Copenhagen, Watkins and Graham made their farewells and Watkins and his friends headed north on a steamer.

In early August, the quartet were welcomed to Angmagssalik by Watkins's Inuit friends. Before leaving for their winter quarters at Lake Fjord, they were joined by Knud Rasmussen, the celebrated Inuit–Danish explorer. A week or so later, while Watkins and Rymill were seal hunting in Lake Fjord, Watkins was nearly swept out of his canoe by waves caused by ice falls from a calving glacier. Luckily, he was unharmed and later that week discussed with his friends whether they could, on the way to Antarctica, stop in India and climb Mount Everest.

On 20 August, Watkins went seal hunting alone. When he did not return, his friends searched desperately for him but found only his waterlogged seal-skin canoe and, on an ice floe, his kayak belt and discarded, soaking wet trousers. It seemed that, as on the previous occasion, blocks of ice had fallen from the glacier face and created huge waves, which had overwhelmed Watkins's kayak. Everyone was distraught, but John Rymill agreed to assume leadership so that Watkins's plans could be completed.

On 5 November, two days before the scheduled memorial service for Watkins, it was announced that Watkins, Rymill, Riley, Chapman and other members of the 1930–31 British Arctic Air Route Expedition had been awarded the Polar Medal. It was the first time the medal had been awarded for Arctic exploration since its inception in 1904.

POLAR POSTSCRIPT: The year after Watkins's death a memorial fund was established in his memory. The Gino Watkins Memorial Fund, administered by the Scott Polar Research Institute and University of Cambridge and the Royal Geographical Society, continues to give grants towards expeditions in polar regions which combine enterprise, scientific research and the prevention of loss of life during expeditions.

Part X

Learning from the Past and Looking to the Future

From childhood, we learn about and from the past through verbal communication, books and other written records, paintings and photographs, natural objects, artefacts and buildings. If we want to know more about a subject (in this case, polar history), we can visit museums, art galleries, archives, sites and buildings in polar regions or join specialist groups who meet in person or communicate through the worldwide web.

The last four objects in this polar history shed light on the links between the 150 or so years covered by this book and the present. Two are buildings, two were once vital pieces or equipment but could now be regarded as artefacts. What they have in common, however, is that their continuing existence is the result of collaboration – whether between scientists and centres of learning, between Inuit and members of other races, between explorers and conservation experts, between clergy in an ancient London church and museum curators thousands of miles away.

As we approach the seventieth anniversary of the Antarctic Treaty, polar scientists from all nations pool their findings and compare their current records with those of the past. Their own and their predecessors' knowledge of ice, climate and polar wildlife – and our willingness to play our own part – are probably our best hope of ensuring that our planet remains habitable for future generations.

We can and should learn from history – and long may we continue to do so. Polar history is not frozen in time: it is dynamic and ever-changing – as evidenced by the fact that, while this book was being prepared for publication, the finding of the wreck of the *Quest* and the publication of a conservation plan for the wreck of the *Endurance* were announced. Who knows what we will learn from polar history over the coming years?

47

'A Polar Centre': The Scott Polar Research Institute Building

The official opening of the new, custom-designed building for the Scott Polar Research Institute (SPRI) in November 1934 marked a turning point for the institute, which had, since its foundation in 1920, been accommodated in a series of Cambridge University buildings. SPRI's new premises were the manifestation of an idea which occurred to the institute's director, Frank Debenham, and fellow geologist Raymond Priestley during the *Terra Nova* expedition. The pair, knowing they needed to consult Priestley's *Nimrod* expedition reports but uncertain where they were stored, devised the idea of a 'Polar Centre', a permanent repository for scientific and other records from their own and other polar expeditions, past, present and future.

The building, on Lensfield Road, Cambridge, was designed by architect Sir Herbert Baker as both modern, functional premises and a memorial to Scott and other members of the South Pole party. The façade includes window keystones with penguin and polar bear motifs, a bust of Robert Scott and a carved inscription, *Quaesivit Arcana Poli Videt Dei*, which, roughly translated, means, 'He sought the secrets of the Pole but saw [those] of God'.[1] Baker's design incorporated a memorial hall, library, archival and other storage facilities, a museum, a gallery and workspaces for researchers; unlike other many Cambridge University buildings, it was open to the public.

The British Polar Exhibition held in London in 1930 had raised awareness of SPRI and its work through displays of polar maps, artefacts, specimens,

Scott Polar Research Institute exterior; image © A. Strathie.

Arctic dome in Memorial Hall; image © and courtesy of Sarah Airriess.

paintings, photographs and the latest version of Herbert Ponting's *Terra Nova* expedition films.[2] *The Polar Book*, published to coincide with the exhibition, also offered a layperson's guide to polar history and the latest scientific developments – not least the isolation of vitamin C, which promised to finally eradicate the 'bugbear scurvy' for those on long voyages and without access to fresh produce.

The new building's architect, Sir Herbert Baker, who had worked for the Commonwealth War Graves Commission, designed the vestibule in the form of a memorial hall with two internal domes. Each dome was decorated with a hand-painted map of a polar region, complete with the geographically appropriate expedition vessels and, in a border, the names of leading explorers. The eye-catching designs had been created and executed by a famous graphic artist MacDonald Gill, his assistant Priscilla Johnston and other helpers.[3]

On 16 November, Gill and Johnston joined over 150 guests for a formal lunch at Gonville & Caius College, where Edward Wilson had studied. Wilson's widow, Oriana, was travelling abroad and unable to attend, but other guests included Scott's widow, Kathleen (who had sculpted the bust over the entrance), and her second husband, Edward Hilton-Young, two of Scott's sisters and Henry Bowers's sister, Lady May Maxwell.[4] The *Terra Nova* expedition cohort included not only Debenham and Priestley, but George Murray Levick, Charles 'Silas' Wright, 'Griff' Taylor and John Mather. Of other expeditions, scientist Louis Bernacchi represented both the *Southern Cross* and *Discovery* expeditions, while *Endurance* expedition veterans included James Wordie and James Marr. Wordie, based on his more recent travels, also ensured the institute's Arctic work was represented, as it was by Denmark's Enjar Mikkelsen, Gino Watkins's friends Freddie Chapman, Augustine Courtauld and Lawrence Wager, Miss Jessie Lefroy (a great-niece of Sir John Franklin) and Isobel Hutchison, an intrepid explorer of Greenland whom Debenham had previously invited to lecture at Cambridge.

During the formal speeches, Cambridge University's Chancellor, ex-Prime Minister Stanley Baldwin, spoke of the new generation of explorers – including Watkins's friend John Rymill and Shackleton's son Edward, who were currently leading expeditions in, respectively, Antarctica and the Arctic.[5] He stressed that as Scott, Edward Wilson and others knew instinctively, short-term goals were not everything. The fine new building would, he said, provide education and training for those continuing to explore 'the partly-known and the unknown'.

Antarctic dome (detail), showing *Discovery*, *Erebus* and *Terror*, *Terra Nova* and *Fram*; image © and courtesy of Caroline Walker.

Antarctic dome (detail), showing *Jane* and *Beaufoy*, *Endurance* and *James Caird*; image © and courtesy of Caroline Walker.

Arctic dome (detail), showing *Erebus*; image © and courtesy of Caroline Walker.

Frank Debenham remained Director of the Scott Polar Research Institute until 1946. Although, during his tenure, polar exploration remained largely a male preserve, the institute describes Kathleen Scott and Oriana Wilson as its 'founding mothers'. Debenham's long-standing assistant, Winifred Drake (initially a volunteer, and later paid), arranged and catalogued much of SPRI's collection.

Debenham's understanding of the history of John Franklin's expedition and family resulted in Jessie Lefroy's bequest to SPRI of a major collection of written material and artefacts previously owned by Jane Franklin and Lefroy's aunt, Sophia Cracroft. Explorer Isobel Hutchison, who may have conversed with Lefroy and May Maxwell when all three attended the opening ceremony in 1934, also left the institute some artefacts collected during her forty years of travel and scientific work in the Arctic region. There are, as Frank Debenham clearly understood, many ways of being involved in and contributing to polar exploration, its history and a polar centre.

48

The *Erebus* Bell

During much of 1845, HMS *Erebus*'s new ship's bell would simply have marked the passage of time and shipboard watches during Sir John Franklin's Arctic expedition. As *Erebus* and *Terror* entered misty, ice-strewn Arctic waters, it might have rung to signal the ships' and men's presence to whalers or local Inuit, or to communicate between the two ships, alert men of danger or recall them to the ships during landings.

The *Erebus* bell was cast in so-called bell metal (or bell bronze), an alloy with high tin content which provides a good ringing tone. The bell bears the British Admiralty's broad arrow mark and is dated 1845, the year the ice-strengthened *Erebus* left London.[1]

In early 1846, while *Erebus* and *Terror* overwintered on Beechey Island, their bells probably rang the customary respectful 'eight bells' in memory of the three men who died during the expedition's first winter. Ship's routines would have resumed as *Erebus* and *Terror* headed south down Peel Sound, but between the writing of the May 1847 Victory Point note (with its optimistic 'All well' ending) and April 1848, it probably tolled at least once a month for John Franklin, Graham Gore and other expedition members who died.

We do not know whether the bell was rung again after Francis Crozier, James Fitzjames and their shipmates deserted the ice-beset *Erebus* and *Terror*. We know that after depositing an updated version of their 1847 note at Victory Point on King William Island, they headed south in the hope of reaching Hudson's Bay Company settlements via the Great Fish River. More graves, remains and relics have been found, but unless and

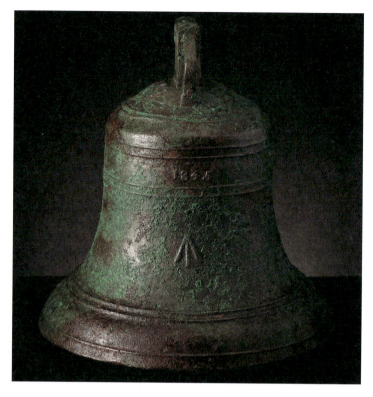

Ship's bell (ref. 89M99X1-1); image © and courtesy of Parks Canada/Marni Wilson, catalogue no. RA4319EF (2017).

Windlass and bell in situ, with archaeologist shining light on bell; image © and courtesy of Parks Canada/Thierry Boyer, catalogue no. 89M00528EF (2014).

Small medicine bottle with 'John Oxley' and 'London' embossed on the glass; Oxley, a well-known pharmacist, was famous for ginger root extract (still used to combat seasickness), but the bottle also contains traces of other remedies and was repurposed as a shot pellet holder. Bottle (ref. 89M99X1-8); image © and courtesy of Parks Canada/Marni Wilson, catalogue no. RA3882EF (2017).

until further notes, letters or logbooks are found, we will not know for sure whether they returned to *Erebus*.

In 2014, following an increasingly close collaboration between Parks Canada and the late Louis Kamookak and other members of the Inuit community living around Gjoa Haven, a wreck was found south of King William Island. Following an intensive six-season search, the latest technology was used to identify the wreck as *Erebus*, but the finding of the wreck – and hence the *Erebus* bell – owned much to Inuit oral history. The wreck lies in around 40ft of water in Wilmot and Crampton Bay. The first items to be raised from the seabed was the *Erebus* bell, which was soon transferred to laboratories for examination, cleaning and conservation. The location of the wreck and that of *Erebus*'s sister vessel (found in 2016, also thanks to Inuit oral testimony) are now designated the Wrecks of HMS Erebus and HMS Terror National Historic Site, which is jointly managed by Parcs Canada and Inuit.

Erebus, despite being located in relatively shallow water, is in a more exposed position than her sister ship so Parcs Canada prioritises dives to her rather than to *Terror*. The *Erebus* bell was the first object to be retrieved, but since 2014 hundreds of objects have been retrieved, examined and conserved.

During nineteenth-century searches for traces of Franklin's expedition, monogrammed cutlery and other objects which could be linked to specific expedition members were found. Objects retrieved from the *Erebus* wreck since 2014 include officers' epaulettes and footwear (with no evidence of names or initials) and communal objects including unmarked tableware and the ship's cannon. Other items including medicine bottles and a daguerreotype polishing plate may have been used by specific individuals or officers in the course of their duties. The finding and identification of the polishing plate raised the intriguing question whether there may be daguerreotypes to be found on the wreck, and, if so, whether any traces of images remain on them.

Another tantalising find is a leatherbound folio of paper, with a quill pen tucked inside the cover.[2] Found in the steward's pantry, only careful and prolonged analysis of it and any other documents on the ship will reveal whether they can tell us more about what happened to Franklin and his men – including when the *Erebus* bell rang for the last time.

49

An Expedition Hut

The hut at Cape Evans in which Scott and his men lived between early 1911 and early 1913 is regularly referred as 'Scott's hut' or the '*Terra Nova* expedition hut'. Although Scott's men were its principal occupants, it and other expedition huts in the Ross Sea area regularly served members of more than one expedition, underlining the frequency and importance of interrelationships between expeditions and explorers. Such exchanges, however, have sometimes provided challenges for New Zealand's Antarctic Heritage Trust (AHT) and others who have, since 1960, sought to conserve the huts and the artefacts in them.

The Cape Evans hut was built with prefabricated panels brought from Britain and erected by a team of men led by ship's carpenter Frankie Davies. Measuring 50ft by 25ft it was, at the time, the largest habitation built in Antarctica. It accommodated Scott's twenty-five-man landing party in relative comfort and included living and sleeping accommodation, scientific and photographic laboratories and an adjacent stable block.

Between 1907 and 1909, during his *Nimrod* expedition, Shackleton occasionally used the *Discovery* expedition hut in McMurdo Sound on his way to and from the Great Ice Barrier. A few years later, members of Scott's *Terra Nova* expedition visited and sometimes stayed in Shackleton's *Nimrod* expedition hut when carrying out scientific work at Cape Royds or, as in Herbert Ponting's case, when photographing and filming its Adélie penguin colony (something Cape Evans lacked). Occasionally, Ponting and others would 'borrow' objects from the Cape Royds hut and bring them back to Cape Evans.

Scott's hut exterior, 2011, showing relics left by occupants over the years; image © A. Strathie.

The so-called 'Tenements' bunks in Scott's hut (as in 2011). They were occupied in 1911 by Bowers, Cherry-Garrard, Oates, Meares and Atkinson, in 1911–13 by Cherry-Garrard and Atkinson (Bowers and Oates died, Meares returned to England) and during Shackleton's *Endurance* expedition by members of the Ross Sea party; image © A. Strathie.

Members of the Ross Sea party of Shackleton's *Endurance* expedition based themselves in the Cape Evans hut after their ship, *Aurora*, was blown off her McMurdo Sound mooring and prevented by ice from returning. In March 1916, after news of the Ross Sea party's predicament was reported in *The Times*, Apsley Cherry-Garrard (one of the last to leave hut in 1913) wrote to *The Times* assuring readers there were sufficient supplies to last for

Herbert Ponting's photograph showing the original occupants of the Tenements: Bowers and Cherry-Garrard (upper and lower left), Oates (centre), and Meares and Atkinson (upper and lower right); vol. I, opp. p. 124, *Scott's Last Expedition*, author's collection.

months.[1] When Shackleton arrived at Cape Evans in early 1917 to relieve the Ross Sea party, he came ashore, but after learning that Arnold Spencer-Smith, Aeneas Mackintosh and Victor Hayward had died, only visited the hut briefly. Before leaving, the remaining members of the Ross Sea party erected a cross in memory of the three men, which still stands on Wind Vane Hill near the hut; an anchor left by *Aurora* is also in the vicinity of the hut.

Members of the 1956–58 Commonwealth Trans-Antarctic Expedition's Ross Sea party (leader, Edmund Hillary) travelled from their base to Cape Evans. When they found the hut full of snow, they attempted to clear it so it would not deteriorate further. They also visited the other McMurdo Sound expedition huts and the Cape Crozier shelter built by Edward Wilson, 'Birdie' Bowers and Apsley Cherry-Garrard during their 1911 winter journey. They brought back from Cape Crozier an abandoned sledge, penguin skins, some of Wilson's drawing pencils, two rolls of unexposed film and some flash powder.

Hillary deposited most of the items with New Zealand museums for safekeeping, but the expedition's photographer George Lowe brought the film rolls and flash powder back to London. While there, he used one film roll which, despite being over forty years out of date, produced satisfactory photographs, including one of Lowe himself and expedition leader Vivian Fuchs beside Scott's statue in Wellington Place, off Pall Mall.[2]

Members of the Trans-Antarctic Expedition raised the issue of the condition of the huts with New Zealand's authorities and 1960 saw the first hut restoration party travelling to Antarctica. Following the establishment of the Antarctic Heritage Trust in 1987, restoration and conservation work was carried out on a more regular basis and by the early 2000s a full conservation programme was under way.

Between 2010 and 2013, the years spanning the centenary of the *Terra Nova* expedition, descendants of Scott and members of his expedition visited the hut. Around this time, AHT conservators found a box containing some twenty badly damaged negatives in Herbert Ponting's photographic laboratory. When the less-damaged ones were printed, it became clear they were taken by a member of Shackleton's Ross Sea party, possibly Arnold Spencer-Smith, who was a keen amateur photographer.[3] Other finds have included a notebook owned and used by *Terra Nova* expedition surgeon and scientist George Murray Levick. AHT policy is that newly discovered artefacts, once restored, are returned to the relevant hut; in this case Levick's notebook joined some 11,000 artefacts already there.

The work of conservation and conservators is never done – and there is always something new to find which might add to the history of polar exploration.

Photograph of (l to r) Vivien Fuchs and George Lowe in front of statue of Captain Scott, Wellington Place, London, taken c.July 1958 using a 1909–10 film roll found at Cape Crozier (ref. rgs 052791(a)); this image © and with permission of the Royal Geographical Society (with IBG); original image © Kodak Ltd, now published in accordance with terms of deposit at the Royal Geographical Society.

50

A Well-Travelled Crow's Nest

The final object in this collection, the crow's nest from Shackleton's *Quest*, was removed from the ship after she returned to London in September 1922. By early 1923, while *Quest* was being refitted in Norway prior to returning to her former role as a sealing vessel, re-fitter Johan Drage removed and stored the purpose-built cabin in which Shackleton had died.[1]

Quest's barrel crow's nest has, for almost a century, been at All Hallows by the Tower church, near London Bridge, from where *Quest* sailed in 1921. In 2022 the crow's nest and *Quest* cabin were reunited at the Shackleton Museum, Athy, Ireland, where, in due course, the cabin will permanently displayed.

By December 1922, Frank Wild and Alexander Macklin had completed the *Quest* expedition narrative and Albert Clavering, an experienced film-maker, had prepared Hubert Wilkins's films for release as *Southward on the Quest*. Interest in both the book and film were high and, to ensure screenings generated funds required to reduce outstanding expedition debts, James 'Scout' Marr assisted with publicity, attending screenings in his kilt or Scout regalia and organising displays in cinema foyers, where audiences enjoyed inspecting the crow's nest and other expedition artefacts.

In early February 1923, Marr publicised the film at the Chelsea Arts Club Ball and spent part of the evening being carried around in the crow's nest.[2] Wilkins's photograph of Wild in the crow's nest was already well known, but when Marr's expedition memoir, *Into the Frozen South*, was published later that year, readers learned that while he sometimes enjoyed being in the 'breezy eminence', at other times he experienced 'a deep sense of loneliness when aloft'.

Quest's crow's nest, in All Hallows by the Tower, near London Bridge, where it has been since 1930; image © A. Strathie.

By early 1930, Wild was living in Africa, Marr was participating in Antarctic expeditions on *Discovery*, and the *Quest* expedition films were seldom shown in cinemas. When the crow's nest, still stored by the film distribution company, was put up for sale, a *Western Morning News* journalist suggested Plymouth Museum should acquire it. The *Quest* had several links to Plymouth: it had not only been her last port of call in Britain, but the expedition's 'generous sponsor' John Quiller Rowett had been educated there before moving to Dulwich College, where he met Shackleton.[3] The suggestion was not taken up, however, and the crow's nest was purchased by Rev. Philip Clayton, incumbent at the ancient church of All Hallows by the Tower, London.[4]

A Well-Travelled Crow's Nest

Frank Wild in *Quest*'s crow's nest; image from Series 6 James Marr collection (ref. Z3qZM8lm7xwo8), courtesy of Library of New South Wales.

Rev. Clayton, better known by his nickname 'Tubby', had been appointed to All Hallows in 1922, shortly before the return of the *Quest* expedition. He had spent much of the Great War near Ypres, where he and a fellow chaplain ran Talbot House, an all-ranks 'Everyman's House' at Poperinghe. In early January 1917, Clayton was delighted when a senior army chaplain, Rev. F.I. Anderson, brought his 'touring' set of Herbert Ponting's Antarctic films for a special Talbot House screening.[5] In October that year, Ponting's fellow Antarctic veterans, Frank Hurley and Hubert Wilkins, visited Vlamertinge, near Poperinghe, where Hurley photographed Wilkins inspecting the graves of British soldiers.[6]

Clayton remained at All Hallows until 1962, combining parish work with raising funds for the church and a network of centres based on the Talbot House model – including donations deposited in the crow's nest, which he redeployed as a huge collecting box. Another of Clayton's projects, the excavation of the church's crypt, resulted in the creation of a museum, where the crow's nest was exhibited when not in use.

During the centenary period of the *Quest* expedition, a three-way collaboration between All Hallows, the Shackleton Museum at Athy and South Georgia Museum in Grytviken resulted in the crow's nest travelling almost as far as it had in 1921–22. In Athy, the crow's nest was reunited with *Quest*'s cabin, which, after being retraced in Norway, had been transported there. Following the exhibition near Shackleton's birthplace, the crow's nest was shipped almost 8,000 miles to Stanley in the Falklands, from where it was transported to South Georgia Museum – which lies close to Shackleton's grave and the memorial cairn that overlooks the harbour where *Quest* was moored in 1922.

The touring exhibition project, three years in the making, demonstrates that, as in exploration, planning and determination reap rewards, both for the collaborating organisations and thousands who saw the crow's nest displayed in places closely associated with Shackleton and his final expedition. As the crow's nest had been a popular fixture at All Hallows for approaching a century, a photogenerated, virtual reality 3-D model of it was created, viewable on site or online. Herbert Ponting and Frank Hurley (who met Clayton in Tehran during the Second World War) were both technophiles and, having pushed at the boundaries of photography in their own Antarctic work, would doubtless have been impressed.[7]

Looking to the future, it is to be hoped that the boundary-pushing *Quest* crow's nest touring project might inspire other holders of expedition artefacts to consider new ways of sharing objects from their own collections in similarly inspired and appropriate ways.

Conclusion

This book has come to an end, but the history of polar exploration continues. Although the newest of the objects featured in *A History of Polar Exploration in 50 Objects* was created almost a century ago, many of them can still be visited or seen in person or by electronic means. For those for whom such objects are accessible, I hope seeing them further enriches your reading. In the meantime, other books (including those cited in this book), films, documentaries, specialist websites or meetings of societies can shed more light on objects, explorers or aspects of polar exploration history which specifically interest you.

For those keen to seek out more objects, visits to explorers' home towns can prove fruitful – a plaque to Edward Parry in the Georgian city of Bath must be one of the most elegant of its kind, while John Franklin and Edward Wilson are well remembered in their home towns in, respectively, Lincolnshire and Gloucestershire. In Norway, traces of Nansen, Amundsen and other explorers are not hard to find, while those visiting Ireland can find tributes to Shackleton in Athy, near Dublin, and to Tom Crean in Annascaul, Co. Kerry. Monuments take many forms, and both John Barrow and Robert Scott are remembered by small-scale lighthouses in, respectively, Ulverston in Cumbria and Cardiff's Roath Park.

Looking forward, however, some of the most enduring legacies of polar expeditions are meteorological, oceanographic, glaciological and other scientific reports which continue to inform scientists up to and including the present day. Accordingly it behoves us to remember not only objects associated with famous 'Polar Stars' but with scientists and others who, like historians, both learn from the past and their predecessors and apply their knowledge to benefit future generations.

Appendix A

Terminology, etc.

Measurements
As in my previous books, measurements, temperatures, units of currency, and so on are as they were used during the timespan of the book. If required, conversion tools can be found online, including those which suggest present-day equivalents of monetary amounts.

Ice
When describing ice and ice features, I have occasionally used explorers' own evocative descriptions, such as 'ice islands' or 'ice barrier'. There are many types of ice (and hence names for them), but the main variants which preoccupied explorers (other than glaciologists) were icebergs (Cook's 'ice-islands'), which can damage ships; pack ice (aka 'the pack'), in which ships can become trapped or 'beset'; and fast ice, which is attached to land and hence safer to travel over than sea ice (the latter being a component of pack ice). A famous example of fast ice is the ice barrier now known as the Ross Ice Shelf. The fact this name was only adopted in the 1950s, over a century after it was charted by James Ross, leads into the next topic, polar place names.

Polar Place Names
The names of polar places and geographical features are a fraught area for polar historians and authors. For example, what is now known as Baffin Bay was previously known as 'Baffin's Bay' and is technically not a bay but a marginal sea or extension of the Arctic and Atlantic Oceans.

Some name changes are widely known (e.g. from New Holland to Australia), but in other cases, endnotes or bracketed explanations are provided. The relatively slow rate of surveying and charting polar areas also resulted in slight name changes over time. For example, King William Land became King William Island following the identification of Rae Strait and it was unclear whether Graham Land was part of an Antarctic peninsula or an archipelago until the 1930s.

The definition of polar regions also varies. The term is generally applied to areas within the relevant circle or beyond 60°N or S, but places such as South Georgia or Southern Greenland, which effectively have polar climates, are sometimes included (and play important roles in polar history).

Caveats regarding changes in place names obviously apply to maps and charts (see Appendix C), which were regularly revised and reissued following voyages of exploration. Maps and the timeline provided (Appendix B) should, when combined with atlases or online facilities, provide adequate detail for most readers.

Appendix B

Summary Timeline

One of the main themes of this book is that leaders of polar expeditions are only one component of a larger whole. As expeditions are often referred to using the names of leaders or their ships, this summary timeline is offered by way of background. Where explorers' names are listed, they are in alphabetical order by surname, with no suggestion of comparative merits. The timeline is also intended to be spoiler-proof for the benefit of readers who prefer *The History of Polar Exploration in 50 Objects* to be a personal voyage of discovery.

1760s–70s

Three circumnavigations under Captain James Cook (1728–79). During the first (on *Endeavour*) and second (*Resolution* and *Adventure*), Cook attempted to delineate and/or reach *Terra Incognita Australis* or the putative Southern Continent. On Cook's third expedition (*Resolution* and *Discovery*), he attempted to traverse the fabled but uncharted Northwest Passage, sailing eastwards from the Bering Strait. In 1773, Constantine Phipps (1744–92) attempted to sail between the Atlantic and Pacific Oceans through an alternative putative route via the North Pole and so-called 'Open Polar Sea'.

1810s–30s

Arctic

The British Admiralty supported several attempts to identify and chart routes between the Atlantic and Pacific Oceans (via either the Northwest Passage or polar regions). Officers involved in these, and later major expeditions, included Francis Crozier (1796–c.1848), John Franklin (1786–1847), Edward Parry (1790–1855), John Richardson (1787–1865), James Ross (1800–62) and John Ross (1777–1856). During this period, naval officers and Hudson's Bay Company employees charted extensive areas of the northern Canadian coastline between the Bering Strait and Northwest Passage archipelago. Attempts were also made to locate the North Magnetic Pole and make magnetic surveys, with a view to fully understanding the effect of the earth's magnetism on mariners' compasses.

Antarctic
Numerous whalers, sealers and naval officers from countries including Britain, Norway, America and Russia worked in, explored and charted islands and land south of Cape Horn and Cape of Good Hope.

1839–43
Francis Crozier and James Ross, on HMS *Erebus* and *Terror*, completed a southern circumnavigation, using charts by Cook and others. Extensive scientific work was carried out, including an attempt to locate the South Magnetic Pole.

1840s–60s
John Franklin was commissioned to command *Erebus* and *Terror* on a voyage to Baffin Bay and the Northwest Passage. Further expeditions followed, with ships entering the Northwest Passage both from Baffin Bay and the Bering Strait.

1872–76
The Royal Society instigated and oversaw a southern circumnavigation on *Challenger*, during which extensive scientific work was carried out, including dredging, which suggested the existence of a sizeable southern continent.

1870s–90s
Arctic
An expedition led by Albert Markham (1841–1918) and George Nares (1831–1915) attempted to reach the North Pole. Norway's Fridtjof Nansen (1861–1930) and America's Robert Peary (1856–1920) both led expeditions to Greenland and attempted to reach the North Pole. Large numbers of mercantile mariners, scientists and privately funded explorers, including Frederick Jackson (1860–1938) and Benjamin Leigh Smith (1828–1913), explored Franz Josef Land, Spitsbergen and other Arctic areas.

Antarctica

Expeditions led by Carsten Borchgrevink (1864–1934), on *Southern Cross*, and by Adrien de Gerlache (1866–1945), on *Belgica*, became the first to overwinter near or below the Antarctic Circle.

1900–20

By now, photography and filmmaking were increasingly the norm.

Arctic

Norway's Roald Amundsen (1872–1928) attempted the first traverse of the Northwest Passage. Amundsen then planned to drift on an ice-bound ship over the North Pole, while Americans Frederick Cook and Robert Peary (1865–1940) attempted to reach it over sea ice.

Antarctica

A series of international geographical conferences resulted in numerous expeditions, including those led by Amundsen (on *Fram*), William Bruce (1867–1921) on *Scotia*, Jean-Baptiste Charcot (1857–1936, *Français, Pourquoi Pas?*), Wilhelm Filchner (1877–1957), Douglas Mawson (1882–1958, *Aurora*), Otto Nordenskjöld (1869–1928, *Antarctica*), Robert Scott (1868–1912, *Discovery, Terra Nova*), Ernest Shackleton (1874–1922, *Nimrod, Endurance*), Nobu Shirase (1861–1946) and Erich von Drygalski (1865–1949).

1920–30s

Although expedition ships remained a cornerstone of polar exploration, the period is distinguished by numerous attempts to overfly polar areas, including by Amundsen, Richard Byrd (1888–1957), Mawson (*Discovery*), Umberto Nobile (1885–1978), Shackleton (*Quest*), Henry 'Gino' Watkins (1907–32) and Herbert Wilkins (1888–1958).

Appendix C

Maps

This book is nominally about fifty objects, but as the text refers to many more expeditions, it is not feasible to provide maps or charts covering each one – and is probably not necessary in an age of online maps and other information (many expedition reports are now online). This appendix includes a selection of historical maps which collectively offer an overview of progress made by explorers and others during the approximately 150 years between Cook's circumnavigations and the beginnings of the aerial exploration era. As can be seen from the timeline in Appendix B, progress in exploring and charting polar regions was not a smooth progression due not only to ice conditions or similar factors, but to war-enforced lulls of varying durations.

For those wanting more detail on Cook's circumnavigations, detailed information can be found on websites including that of the Captain Cook Memorial Museum, Whitby. Recent books by Peter Fretwell and the late Patrick Quilty (see Bibliography) offer maps of the Antarctic in profusion. As to the Northwest Passage, maps from the 1939 edition of Cyriax (see Bibliography) are included below in this section. Details of sources of maps are in the Bibliography.

Maps

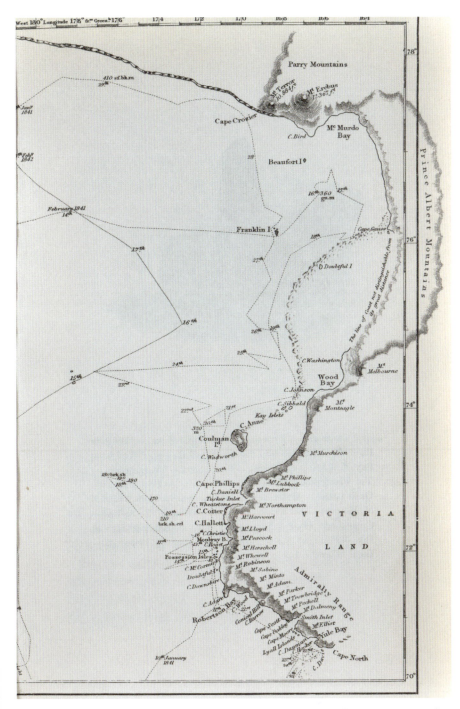

James Clark Ross's 1841 chart of Victoria Land, showing McMurdo Bay (now Sound), Mounts Erebus and Terror and the western end of the ice barrier (now Ross Ice Shelf); from Ross, James Clark (1847).

Map showing the Northwest Passage region as known in 1845, when Franklin set sail; from Cyriax (1939 edition).

Maps

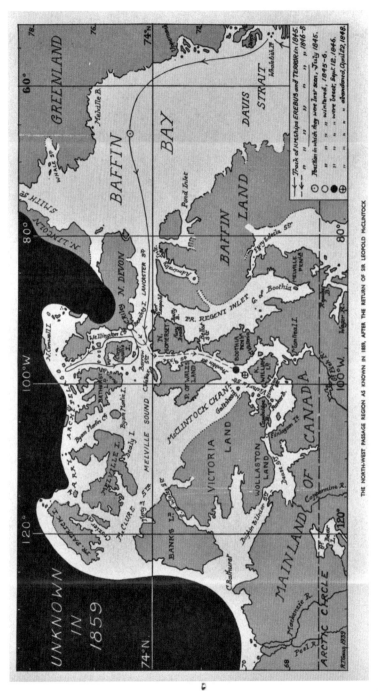

Map showing the Northwest Passage region as known in 1859 after return of McClintock's search expedition; from Cyriax (1939 edition).

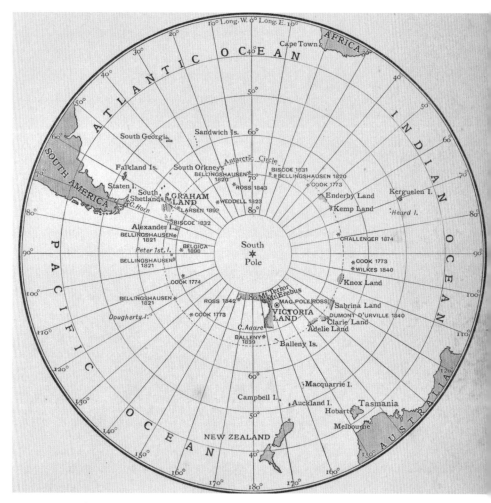

Map showing the extent of charting of Antarctica at the end of the nineteenth century; from Scott, Robert F. et al. (1905).

Map showing the route (in red) of Amundsen's voyage through the Northwest Passage; from Amundsen (1908).

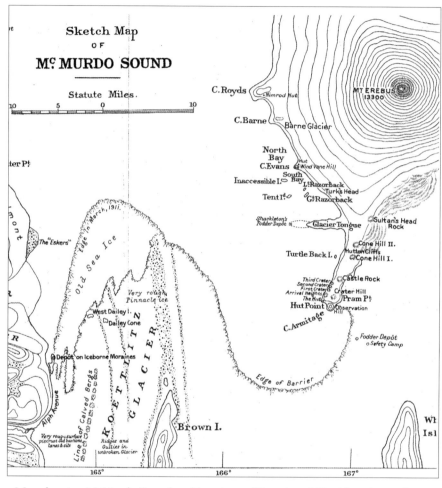

Map showing McMurdo Sound and locations of Scott's and Shackleton's expedition huts; from Scott, Robert F. et al. (1913).

Maps

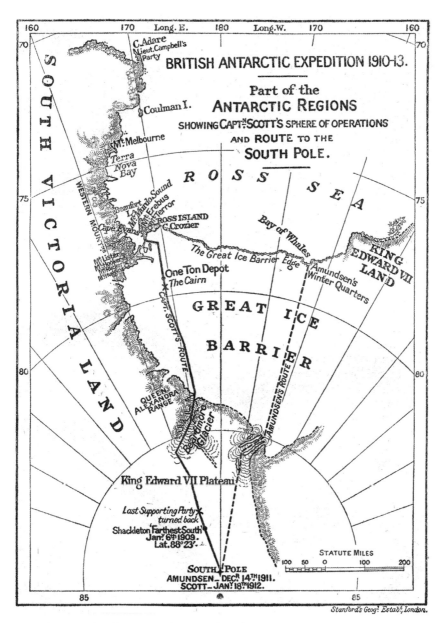

Map showing Shackleton's Farthest South and Amundsen's and Scott's routes to the South Pole; from Ponting (1920).

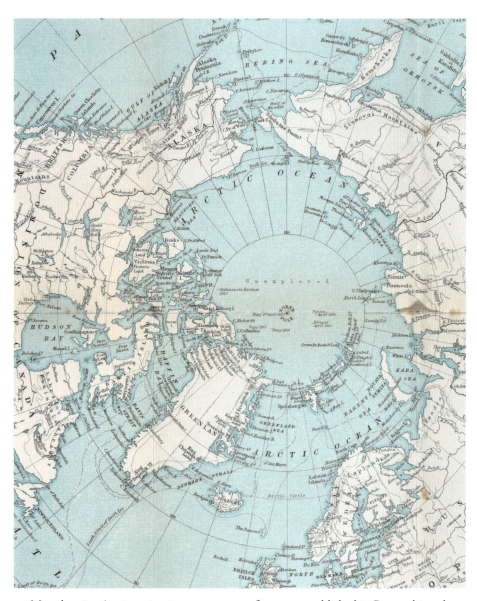

Map showing Arctic region in *c.*1930; extract from map published in Bernacchi et al.

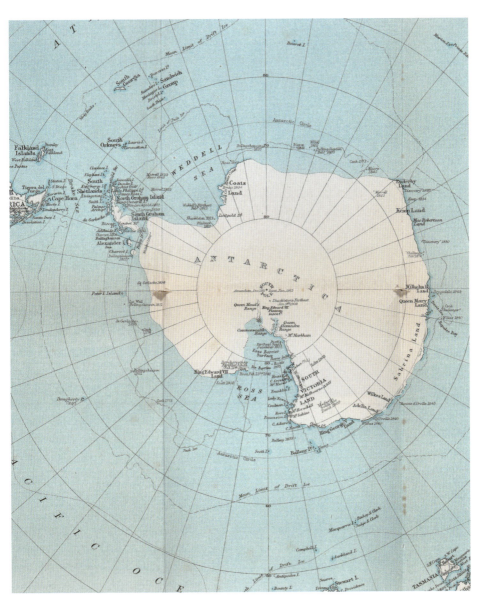

Map showing Antarctic region in *c.*1930; extract from map published in Bernacchi et al.

Acknowledgements

All books are collaborative efforts and the nature and range of *A History of Polar Exploration in 50 Objects* makes this the most collaborative of the four polar-related books I have researched and written over the past fifteen years. All four have been published by The History Press and I thank Gareth Swain, Mark Beynon, Katie Beard, Jezz Palmer, Chrissy McMorris, Cynthia Hamilton, Claire Hartley and their colleagues for commissioning and producing this, my most 'lavishly illustrated' book, with such care.

I am also grateful to the numerous museums, galleries, archives, other organisations and individuals who have facilitated my research and provided images and permissions for this book. In Britain and Northern Ireland (location given unless obvious from name), organisations include the Captain Cook Memorial Museum, Whitby (Maria Aparicio and colleagues); Derbyshire Record Office, Matlock (Sarah Chubb and colleagues); Discovery Point, Dundee; Dominic Winter Auctioneers, Cirencester (Chris Albury); George Waterston Memorial Centre & Museum, Fair Isle (Anne Sinclair); Hull Culture and Leisure (Robin Diaper, Susan Capes); The Hunterian, University of Glasgow (Neil Clark and colleagues); National Maritime Museum/Royal Museums Greenwich, London (Jeremy Michell, Claire Warrior); National Portrait Gallery, London; Natural History Museum, London (Stephen Atkinson); National Library of Scotland, Edinburgh (Paula Williams and colleagues); National Museums Northern Ireland, Belfast; North West Castle Hotel, Stranraer (Bespoke Hotels); Royal Geographical Society with IBG, London (Eugene Rae, Joy Wheeler and colleagues); Scarborough Museums Trust (Jim Middleton); Science Museum/National Science & Media Museum, London and Bradford; Scott Polar Research Institute, Cambridge (Lucy Martin, Naomi Bonham); Shetland Museum and Archives, Lerwick (Carol Christiansen); Spink & Son, London (Marcus Budgen); Stranraer Museum; Stromness Museum and Archives, Orkney (Janette Park and colleagues); The Poetry Society, London (Judith Palmer); Topfoto Image Archives, Kent (Millie Moore and colleagues);

United Kingdom Antarctic Heritage Trust, Cambridge (Camilla Nichol and colleagues); University of St Andrews Libraries and Museums (Lydia Heeley and colleagues); Wellcome Collection, London; Whitby Museum (Hazel Wright and colleagues); The Wilson Art Gallery and Museum and The Cheltenham Trust (Philippa Turner and colleagues).

Like polar exploration, this book has been an international undertaking, and I also thank the Antarctic Heritage Trust, Christchurch, New Zealand (Lizzie Meek and colleagues); Art Gallery of Ontario, Toronto (Alexandra Cousins); ASA *Uruguay*, Buenos Aires; Australian Museum, Sydney; Berkshire Museum, Pittsfield Mass., USA (Jason Vivori and colleagues); Biblioteca Nacional de Chile, Santiago (Gastón Carreño González); Canterbury Museum, Christchurch, New Zealand (Sarah Murray, Nicolas Boigelot and colleagues); Fram Museum, Oslo (Geir Kløver); Hinkler Hall of Aviation (Tracey Kelly and colleagues) and Hinkler House (J. A. 'Lex' Renold), Bundaberg, Australia; Historic Dockside Museum, Stanley, Falkland Islands; Library of Congress, Washington DC; Metropolitan Museum of Art, New York; Museo Fin del Mondo, Ushuaia, Argentina; Museo Maritimo e del Praesidio, Ushuaia; Museums Victoria, Melbourne; Orsi Libri Rare Books, Milan; the Parks Canada Underwater Archaeology Team, Ottawa (John Ratcliffe); Port Lockroy museum, Goudier Island, Antarctica; Queen Victoria Museum and Art Gallery, Launceston, Tasmania (Andrew Parsons and colleagues); Roald Amundsen's House, Svartskog, and MiA/Museums in Akershus, near Oslo (Anders Bache); South Australian Museum, Adelaide (Mark Pharaoh); South Georgia Museum, Grytviken (Jayne Pierce and colleagues); State Library of New South Wales, Sydney (Dixson and Mitchell Libraries); State Library Victoria, Melbourne; Talbot House, Belgium; Tasmania Museum and Art Gallery, Hobart (Dr Mary Knights, Jacqui Ward and colleagues); Waitaki Boys High School, Oamaru, New Zealand (Rector Darryl Paterson).

Many thanks also go to the numerous individuals who have assisted with research, illustrations, permissions, advice and encouragement. Of these, special thanks go to Sophie Wilson, whose patient proof-reading and sound advice ensured my text reached fruition. I am also grateful to relatives and descendants of explorers including Dr David Wilson, Philippa Wordie, Philippa Foster-Back, Judy Skelton, Susannah Ferrar and Janet Crawford. Thanks also to Sarah Airriess, Paul Baker, Sarah Baxter (Society of Authors), Brad Borkan, Mensun Bound and Joanna Yellowlees-Bound, Helen Brown, Rachael Carr, Jan Chojecki, Dr Jean de Pomereu, John Dudeney, Gill Fargher, Richard Fattorini, Paul Firth, Anne Fletcher; Julia Fortes, Ann-Rachael Harwood, Allegra Huston, Jane Kang, Jan-Emil Kristoffersen, Heather Lane, Stuart Leggatt (Meridian Rare Books), Louis Levitz, Mme Anne Longépé and Charlotte Service-Longépé, Ken McGoogan, Colin Monteath, Esther Morgan, Shane Murphy, Alan Noake, Dr Russell Potter, Michael Pritchard (Royal Photographic Society), Beau Riffenburgh, Eva-Linn Röjerstrand, Allegra Rosenberg, Michael Rosove, Chet Ross (Chet Ross Rare Books), Dr Stephen Ross, Stephen Scott-Fawcett, Michael Smith, Glenn Stein, Fabiënne Tetteroo and Caroline Walker. The names of other fellow writers whose books have informed mine appear in source notes and the Bibliography.

While researching for this book, I visited both polar regions and thank those who helped make these voyages of discovery both useful and enjoyable: Angie Butler and Caro Mantella (Ice Tracks), Sarah Scriver and team (Polar Latitudes), Susan Adie and team (Aurora Expeditions); fellow travellers from both voyages, including in particular a group of *Island Sky* shipmates who, thanks to fellow writer Brad Borkan, continue to meet regularly.

Like most non-fiction writers and researchers, I enjoy delving in archives, but also regularly use Ancestry.co.uk, British Newspaper Archives and other online databases and resources. While all play their part, I am particularly grateful to Steve Scott-Fawcett and Russell Potter who, respectively, 'chair' the excellent Sir Ernest H. Shackleton Appreciation Society and Remembering the Franklin Expedition Facebook groups and encourage members to collaborate and share research for the greater good. I have already acknowledged fellow authors, but also owe a debt to booksellers, publishers, auctioneers and others who make rare and hard-to-find books accessible, including the Erskine Press, Francis Edwards, Glacier Books, Hay Cinema Bookshop, Kingsbridge Books, Meridian Rare Books, Gloucestershire Libraries and (online) Project Gutenberg and HathiTrust.

I warmly thank family members, friends and others I have encountered on my fifteen-year personal 'polar journey' for their interest in and encouragement of my research and writing, for buying and reading my books, for attending talks and events, for drawing my attention to material relevant to my research – and for continuing to send me penguin-themed missives, which never fail to cheer. Your continuing support is much appreciated.

As is customary, I apologise for any omissions from the above or from acknowledgements elsewhere in the book – this does not imply a lack of gratitude. I also apologise for any inadvertent errors in the book; these are of my own making and no reflection on those named above.

Last but certainly not least, I thank readers of my three previous books and hope that they and others will enjoy reading or dipping into *A History of Polar Exploration in 50 Objects*.

Notes

As this book provides an overview rather than a detailed history, endnotes are provided mainly when information is not easily found in multiple sources or the public domain. As the bibliography is long and diverse, the endnotes for each object begin with a 'specific sources' section with short-form titles (surname and bracketed information when required for clarity) of publications relating to that object.

The following abbreviations are used in endnotes, captions, etc.:

BNA – British Newspaper Archive
DRO – Derbyshire Record Office
NMM – National Maritime Museum
NPG – National Portrait Gallery
RGS – Royal Geographical Society
RMG – Royal Museums Greenwich
SLNSW – State Library of New South Wales
SPRI – Scott Polar Research Institute

Part I: Laying Foundations
1. Cook's observations were published in a Royal Society report in April 1767.
2. Lieutenant was the lowest rank at which officers could command voyages of such duration or importance.
3. Palliser sailed with Scottish naval doctor James Lind, so he would have known of the latter's work on anti-scorbutics.

1: HMS *Resolution* in Pack Ice
Specific sources: Beaglehole; Hough; Hamilton; Robson; Cook; Captain Cook Memorial Museum (Whitby); Whitby Museum; Captain Cook Society website; RMG and BNA websites.
1. According to John Robson, neither Cook nor Palliser suggested Whitby cargo vessels to the Admiralty.
2. Aurorae Borealis were recognised from the seventeenth century.
3. The last, much-quoted phrase is from Cook's journal, 30 January 1774.
4. Dutchman Jacob Roggeveen visited Easter Island on Easter Sunday, 1722.
5. The island was first visited and named (for himself) by Londoner De la Roché in the 1670s; Spanish mariners called it Isla San Pedro (from sighting it on St Peter's Day); Cook's designation was later simplified to South Georgia.

2: Elizabeth Cook's Ditty Box
Specific sources: As per Chapter 1; State Library of New South Wales, Sydney.
1. It is unclear exactly how Cook met Elizabeth. The consensus appears to be that Cook first met her when her father became Cook's London landlord; some suggest that Cook attended Elizabeth's christening, possibly in the capacity of godfather.
2. Constantine Phipps, commanding two Admiralty bomb ships, reached 80°N before becoming ice-beset.
3. *Caledonian Mercury*, 7 December 1776.
4. *Tapa* fabric is made from bark. Unfinished panels from the waistcoat are in the Mitchell Library, SLNSW (Call No. R198, ref. 421794).
5. Initial newspaper reports are dated 1 January 1780 (BNA); Cook's and other reports reached the Admiralty in late January. Cook used 'Owhyhee' rather than 'Hawaii'.
6. Clerke died from tuberculosis contracted before leaving England.
7. During Cook's stay, islanders commemorated Lono, a demi-god. Some accounts suggest islanders may (possibly due to Cook's height) have thought Cook was a demi-god (e.g., 'Life and death of Captain James Cook as the Hawaiian god "Lono"', Lars-Benja Braasch, Munich, 2005 paper, pub. 2009, reprint available online).
8. Elizabeth left the ditty box to family members; after it passed from the family, a subsequent owner donated it to SLNSW.
9. Young relatives of naval officers regularly sailed with them or had their names entered in muster rolls (which later gave them seniority even without them sailing). The 1870 Great Hurricane death toll remains a record for an Atlantic hurricane.
10. Some newspaper reports suggested James's death involved foul play (unproven); he was buried alongside his brother Hugh in St Andrew the Great, Cambridge.
11. British Museum, ref. M.4837.

3: Scoresby's Barrel Crow's Nest

Specific sources: Stamp (ch. 7); Scoresby and Jackson (intro. *Arctic Whaling Journals*, vol. III); Whitby Museum.

1 Scoresby's biography of his father notes the parallels between his father's and Cook's origins and early careers.
2 Whale oil was used for lamps, candles, soap and paints; whale ribs were used for umbrella spokes and women's corset stays.
3 Cook notes New Zealand was sighted two hours earlier from the top mast than from the deck.
4 Scoresby briefly served under William Bligh, whom he suggests was a harsh taskmaster.
5 William Scoresby (senior) had previously given Banks polar bear furs for his wife.
6 For letters, etc., see Stamp, pp. 66–70.
7 The Admiralty's Second Secretary was an administrator but, like other senior civil servants, had considerable influence.
8 Scoresby, the Stamps and others depict the relationship between Scoresby and Barrow as adversarial, but Jackson (in the introduction to Scoresby, *Arctic Whaling Journals*, Vol. III) suggests they simply had incompatible goals and objectives.
9 Although Parry's full name was William Edward, he generally used Edward.

4: A Panorama of Spitsbergen

Specific sources: Fleming; Ross, M.J.; Dodge; David; Potter (2007); Lewis-Jones; O'Dochartaigh; Plunkett (paper); Ralph Hyde's 'Dictionary of Panoramists …' (online); The Regency Redingote blog and other online articles on Barker's and Marshall's panoramas; Buchan's and John Ross's expedition reports (online); BNA newspaper advertisements for 1820 panoramas.

1 During the First World War, Britain adopted the less-Germanic spelling 'Spitsbergen'. Spitsbergen originally referred to the archipelago, rather than, as now, the largest island of the Svalbard archipelago.
2 Barker's wife was a daughter of William Bligh, who served in the Arctic with Cook.
3 George Cruikshank's cartoon was particularly scathing.
4 From Marshall's description, these drawings are not those used by Barker and may include some of those which appear in Beechey's own expedition account (published 1843).
5 *Bath Chronicle*, inc. 16, 23 January 1823.

5: William Scoresby's Manuscript

Specific sources: As per Chapter 3 above; Whitby Museum website and catalogue for 'Scoresby's Arctic' exhibition (2022).

1 Robert Jameson, *A System of Mineralogy* (1816).
2 *Hecla* and several naval vessels used in the Arctic were almost double that tonnage.
3 From Scoresby's 1820 *Arctic Whaling Journals*.
4 Thanks to Helen Brown for loaning her late father's copy of Scoresby's paper.

Part II: Exploring North and South
1 The islands lay at a similar latitude south as the Shetland Islands in the north.
2 Smith and Bransfield passed but did not chart Deception Island; other remote islands included the South Orkneys and Elephant Island. Bransfield may have named Trinity Land for the London naval establishment Trinity House.
3 Von Bellingshausen was sufficiently near the Antarctic coast for that to be feasible.

6: Weddell's 'Sea Leopard of South Orkneys'
Specific sources: Weddell; Bulkeley; Jones, A.G.E. (1982); Quilty; Rubin (paper); Basberg & Headland (paper).
1 Whalers and sealers did not carry sufficient provisions for long, speculative journeys.
2 *Leptonychotes* refers to the narrow claws on Weddell seals' flippers.
3 Enderby's Land was also seen by Weddell and Bellingshausen; Graham's Land was probably the first charted part of the Antarctic mainland.

7: Edward Parry's Deck Watch
Specific sources: Fleming; Parry; Ross, M.J.; Clockmakers' Museum/Science Museum.
1 Admiralty equipment could be custom-made or provided from stock.
2 In his 1820 book, Scoresby suggested using Inuit-style dog sledges to cross sea ice, but as Barrow disliked both Scoresby and John Ross, he generally ignored their suggestions.

8: A Canister of Meat
Specific sources: Fleming; Dodge; Ross, M.J.; Rotunda Museum, Scarborough.
1 Johnstone, who was Yorkshire's Member of Parliament until 1832, appears to have been related to the Earls of Annandale who owned land in Dumfriesshire, John Ross's home county.
2 Booth's contribution exceeded £10,000; Ross invested £3,000 from his own funds.
3 Ross's charts were later amended to show a strait through the isthmus.
4 It is unclear whether the Inuit knew of what was charted in the 1850s as the Bellot Strait.
5 Magnetic poles move constantly so are not fixed points; John Ross later claimed that James Ross did not tell him he was heading for the Magnetic Pole.
6 Somerset House was named either for the Admiralty's London headquarters or North Somerset/Somerset Island.
7 Batty Bay lies between Fury Beach and Barrow Strait.
8 Now known as Back('s) River.
9 The can Ross gave to the son of his friend, the Earl of Stair, was opened (per Ross's instructions) on 4 August 1869, the recipient's twenty-first birthday; the can is in Stranraer Museum (ref. WIWMS1964).

9: James Ross's Career-Defining Portrait
Specific sources: Dodge; Ross, M.J.; RMG/NMM and NPG (portraits); Potter (2007 and paper); BNA (advertisements for and articles on exhibitions, etc.).
1 Ross, M.J., *Polar Pioneers*, p. 191.
2 Newspapers used his full name from 1829.

3 It was claimed there was no precedent for splitting the money.
4 Parry's portrait (Samuel Drummond, *c*.1819) is in the NPG (ref. 5053).
5 *True Sun*, 17 January 1834; Green's portrait of John Ross is in the NPG (ref. 314).
6 *Atlas*, 26 January 1834.
7 *Atlas*, 30 March 1834; *Morning Post*, 2 April 1834.
8 *Morning Advertiser*, 24 March 1824. Hawkins's portrait of John Ross is RMG/NMM ref. BCH2983.
9 ArtUK website ('Art Detective: John Ross portraits') cites Green (No. 330), Faulkner (No. 261) et al.
10 *London Evening Standard*, 31 May 1834.
11 I have used 'Anne' in accordance with census returns.
12 Rhymer-Jones edited sections of Ross's expedition report.
13 Ross added an extension which accommodated a large-scale representation of *Victory*'s cabin and at least one panorama mural; he also added a camera obscura, stone-carved coat of arms and a private slipway.

10: Rossbank Magnetic Observatory, Hobart
Specific sources: Ross, James; Ross, M.J.; Dodge; Fleming; Palin; Smith (2006); Savours & McConnell (paper).
1 Although Van Diemen's Land formally became Tasmania in 1856, the names were used concurrently.
2 Bock was convicted in 1823 of administering 'concoctions of certain herbs' to his pregnant mistress to facilitate a miscarriage.
3 Smith, *Icebound*, Chapter 11.
4 D'Urville named land and a species of penguins Adélie for his wife, Adèle; '*détermination*' as used by d'Urville suggests 'establishing the location of' rather than reaching the Magnetic Pole.
5 Ross mentions the Coulman-related names and subsequent marriage to Anne Coulman in his published expedition report.

11: A Great Icy Barrier
Specific sources: As per Chapter 10.
1 The Great Ice Barrier was sometimes known later as the Ross/Ross's Barrier (e.g. during Scott's expeditions) but was only renamed the Ross Ice Shelf in the 1950s.
2 Ross's Parry Mountains were, like John Ross's Croker's Mountains, illusory refractions.
3 Ross, like Cook, uses eastern longitudes in excess of 180°E rather than switching during a voyage to west-based longitudes.
4 Ross believed Wilkes saw misleading refractions but considered d'Urville and Balleny's findings reasonably accurate.
5 Ross and Crozier's stay in the Falkland Islands is recalled in street names in Port Stanley (which later replaced Port Louis as the islands' main port).
6 Ross, like Cook, believed anti-scorbutics and regular rewatering and reprovisioning of ships kept men healthy.

12: Francis Crozier's Penguin
Specific sources: As per Chapter 10 above; Finnegan (paper); Rosove (paper); Belfast Natural History & Philosophical Society website.
1 After the initial curator left, the extent of the gifts from Crozier and others resulted in display space being expanded.

Part III: The Northwest Passage: The Search Continues
1 Ross, before marrying, undertook not to accept any more long-distance commissions. Battersby and others suggest Barrow's championing of Fitzjames was in recognition of his assistance to Barrow's son, George, while both were in Asia.
2 Magnetism readings supplemented those by North American magnetic stations; Crozier, an acknowledged expert in magnetism, was unhappy about Fitzjames's additional role.
3 Potter et al., Letter 85, p. 120.

13: A Daguerreotype
Specific sources: Potter et al. (originals of Goodsir's letters are with the Royal Scottish Geographical Society, ARC.4.3/2); Cook (paper); Kaufman (paper); Pritchard; Batchen.
1 Scottish medical training included more natural science courses than its English equivalent.
2 The sisters' father, Austen Lefroy, was a cousin of John Rae's mentor, Henry Lefroy.
3 Adamson used William F. Talbot's calotype process.
4 Potter et al., Letter 90.
5 Royal Society blog, 'King Lion', April 2020.
6 Franklin said initially he had no room for scientists (Franklin to Richardson, 22 February 1845, DRO, D8760/F/FJR/1/1/90). Harry was promoted after Prince Albert 'suggested' to Lord Haddington of the Admiralty that all 'voyages of discovery' should include a naturalist.
7 There is no mention in Harry's letters of his being invited.
8 Richardson's second wife, Mary Booth, died in April 1845.
9 When Goodsir's reports were published in England, Joseph Hooker wrote to *Annals and Magazine of National History* (Vol. 16, pp. 238–39) to point out that he and James Ross had, between them, already disproved Forbes's theory by finding invertebrates below 300 fathoms in both polar regions.
10 Potter et al., Letters 166 and 167.
11 From list of exhibits on pp. 410–16 in *The Scottish Geographical Magazine*, vol. XI (1895); the portrait is described (p. 415, item VII) as 'Portrait of Dr. Harry Goodsir taken the day before he sailed in the "Erebus"'; the use of the word 'taken' suggests it may be an original or copy Daguerreotype.
12 See 'Visions of the North' blog (Potter), 26 May 2023.

14: A Rock at Port Leopold

Specific sources: Ross, M.J. ; Dodge, Fleming; Jones, A.G.E. (paper); Burford; Ross, James Clark, with others (1850).

1 John Ross, by then 70, offered to lead a relief expedition, but his suggestion was turned down.
2 The couple's collective nickname for Francis 'Frank' Crozier and Franklin.
3 Ross, M.J., *Polar Pioneers*, pp. 298–99.
4 William Baillie-Hamilton (Barrow's successor), Edward Parry and whaling master Thomas Ward judged claims for Jane Franklin's rewards.
5 Later charted as Peel Sound.
6 Some tinned provisions were apparently below the stated weights.

15: John Rae's Octant

Specific sources: Richards; Lefroy; Rae/intro. McGoogan (both books); John Rae Society website; Stromness Museum and Archives; Orcadian historian Tom Muir's online posts on John Rae; Claydon, p. 144 and note 31 (friendship of John Richardson and Rae's father).

1 Lefroy previously established the St Helena magnetic observatory as part of James Ross's southern hemisphere magnetic survey.
2 Ross's party including Orcadians, a Shetlander, a Scottish highlander, two *voyageurs* and Cree and Métis guides.
3 The Inuit remembered the Rosses' 1829–33 expedition.
4 The marks were used to identify and deter stealing of Admiralty property.
5 Rae asked Simpson to forward material to the RGS (Rae to Simpson, 10 June 1851, Rae/McGoogan, 2014).

16: A Graveyard on Beechey Island

Specific sources: Alison Freebairn's 'Fingerpost' blog ('Robert Goodsir and the Franklin Graves on Beechey Island' and others); Logan Zachary's Illuminator blog ('The Cape Riley Rake').

1 Harry's other younger brother, Archibald, died in 1849 while Robert was on his first search with Penny.
2 Earlier cans (supplied by Donkins) apparently caused no ill effects, but some suggest that the contents of Goldner's cans, due to them being hastily welded, went off and caused stomach upsets or worse.

17: Eleanor Gell's 'Franklin Search' Collection

Specific sources: DRO, including albums and correspondence as cited; Potter et al.; Owen; Beattie & Geiger; McGoogan (three books).

1 Potter et al., Letter 149.
2 *Ibid.*, Letters 180 and 184 (the former refers to her letter to Crozier).
3 Eliza Peddie (wife of *Terror*'s surgeon) to Eleanor Gell, 20 February [1852], DRO: D8760/F/FEG/1/52/71.
4 In summary, should Franklin be declared dead, his will would be proved, and funds bequeathed to Eleanor by her mother released to her (something of which Jane Franklin, who had read her husband's will, was aware).

5 The Admiralty now needed ships, men and funds to support Britain's participation in the Crimean War.
6 *Illustrated London News*, 28 October 1854.
7 Letters held at DRO suggest that Eleanor's Franklin relatives, including Sophy's mother (known as 'Aunt Cracroft') and sister, Emma Lefroy, sometimes acted as peacemakers within the family.
8 The new Canterbury diocese was never established.
9 See British North Atlantic Telegraph Expedition; Rae letters are at DRO/D8760/F/GJP/10/1-3.

Part IV: A New Start
1 Millais, John Guille, *The Life and Letters of Sir John Everett Millais* (1899), Vol. II, Ch. XIV.

18: A Photograph of Antarctic Icebergs
Specific sources: Jones, Erika; Thomson et al., Vol. 1 (Expedition Narrative), Chapters X and XI (icebergs); Brunton; Gould (paper); Hood (paper).
1 No photographs from James Ross or Franklin's expeditions have been traced (Ross may have taken a camera, Franklin took a daguerreotype apparatus).
2 After Nottingham-born Newbold left the ship in South Africa, he appears to have remained there until around 1909–10 before moving to Illinois, USA, where he died in February 1911 (*Zion City Independent*, 10 February 1911).
3 It is unclear whether Newbold, Hodgeson or both took photographs in South Africa.
4 See Thomson et al. for iceberg descriptions and illustrations (Chapters X and XI).
5 The reasons for the omission are unclear, but many stereoview companies did not credit their photographers.
6 See Stein, G.M., 'Photography Comes to the Polar Regions' in *Antarctic*, Vol. 26, No. 1.

19: A Menu for a Banquet
Specific sources: *Hampshire Advertiser*, 2 December 1876 et al. (BNA); Nares; Markham (1878).
1 The front page of *Illustrated London News*, 5 June 1875, shows the ships departing from Southsea.
2 Some Franklin relief and rescue expeditions charted Arctic areas.
3 Lime juice rations apparently lost their anti-scorbutic qualities, and the continuation of customary alcohol rations may have exacerbated the situation; as indicated on James Ross's expedition, heavy sledging appears to have accelerated some cases.
4 Markham's sledge flag, made for him by Mrs McClintock, is at NMM/RMG.

20: Cornelius Hulott's *Resolute* Box
Specific sources: Stein; family information including from Ancestry.co.uk, The National Archives, Lynne Flatman and other family members; BNA newspaper reports featuring Hulott.
1 The design appeared in *The Graphic*, 20 September 1879.
2 Approximately 500lb of meat rotted due to damaged/faulty cans.

3   Armstrong was a near contemporary of Harry Goodsir at Edinburgh University. The doctors wondered why Hulott, despite his slight build, appeared to have more stamina than others; some (e.g. Stein, p. 143) suggest Hulott needed fewer calories but received equal rations, but others suggest he declined his rum rations.
4   For more on *Investigator*'s sledge journeys, see Stein, Chapter 9 and Appendix 5.
5   Franklin by then was also described as 'Discoverer of the Northwest Passage'.
6   *Morning Post*, 6 February 1899.

21: 'On Board *Eira*': From the '*Eira* 1880' Album
Specific sources: Capelotti; Conan Doyle.
1   Leigh Smith and his sister Barbara (aka Bodichon, an artist and women's rights campaigner) were out-of-wedlock children of Benjamin Smith, a radical Member of Parliament; his birth status appears to have made him shy of publicity.
2   Markham may have suggested that Northbrook and Bruce Islands were named for, respectively, immediate past and current RGS presidents; it is unclear whether Cape Flora was named for a family member or the flowers found there.
3   RGS *Proceedings*, Vol. 3, March 1881, pp. 129–50 (1880 expedition) and Vol. 5, No. 3 (April 1883), pp. 204–28 (1881–82 expedition).
4   The Admiralty, rather than provide a ship, donated £5,000; around £14,000 of rescue costs were funded by the RGS and individuals.
5   *St James Gazette*, 22 August 1882.

22: *Illustrated London News* Front Page
Specific sources: Jackson; Nansen; Huntford (1997); Fram Museum information; BNA newspaper reports on Jackson's expedition.
1   Nansen and others assumed that a transpolar current, passing near the North Pole, transported wreckage from USS *Jeannette* from north of Siberia to Greenland.
2   Leigh Smith's charts only showed capes and tentative outlines.
3   Bruce participated in 1892–93's Dundee whaling expedition.
4   Harmsworth licensed the photograph to the *Illustrated London News* and others.
5   Thanks to Federico Orsi of Orsi Libri for permission to quote the inscription.

Part V: Antarctica Revealed
1   Murray had, in 1893, submitted a paper to the RGS calling for a revival of Antarctic exploration (RGS *Proceedings*, January 1894).

23: A Stereoview of Adrien de Gerlache and a Weddell Seal
Specific sources: Amundsen, Roald, et al., *Belgica Diary* (1999); de Gerlache; Lecointe; Sancton.
1   It was only in 1899 that Markham secured sufficient funds to commission *Discovery*.
2   Cook, a late recruit as ship's doctor, had travelled to the Arctic with fellow American Robert Peary.
3   Plans appear to have been regularly adapted between 1895 and *Belgica*'s departure.
4   Causes of delays following arrival in Punta Arenas (1 December 1897) also include crew issues, Cook's anthropological studies in Tierra del Fuego, the grounding

of *Belgica*, and inaccurate charts suggesting a non-existent channel between west Graham Land and the Weddell Sea.
5 Amundsen (as above), 20 July 1898.
6 Amundsen, at his parents' behest, briefly studied medicine.
7 Cook uses the term 'polar anaemia', but Guly (paper) suggests this is closer to beri-beri than scurvy; Sancton suggests Cook's photographic chemicals could have affected men's health.

24: Louis Bernacchi's Cape Adare Home
Specific sources: Crawford; site visit.
1 *Southern Cross* (as *Pollux*) was designed as a whaler by Colin Archer, who designed Nansen's *Fram*.
2 Bernacchi, born in Belgium to an Italian family, was raised in Britain and Tasmania, then moved to Melbourne to study magnetism. Borchgrevink, who also lived in Australia, failed to raise sufficient funds for his expedition, so travelled to London to publicise his plans during the 6th International Geographical Congress.
3 Norway's separate constitution was signed in 1814.
4 Bernacchi cites Borchgrevink's generally erratic behaviour, regular changes of the men's duties and failure to make best use of Colbeck's and his magnetic training.
5 It is suggested that Hanson's death may have resulted, at least in part, from a disease which affected several crew members following a reprovisioning stop in the South Atlantic.
6 Borchgrevink, Colbeck, Bernacchi and Evans reached approximately 78° 50'S.
7 Markham resorted to the long title after Newnes referred to Borchgrevink's expedition as the 'British Antarctic Expedition'.

25: RRS *Discovery*
Specific sources: Savours; Scott, Robert F. (1905); Skelton & Wilson; Skelton.
1 Savours, A., *The Voyages of the Discovery* (2013), Chapter 2.
2 Markham first met Scott when the latter was attending a naval training establishment run by Albert Markham; Markham would also have known George Egerton, Scott's mentor and ex-commanding officer, through Nares's Arctic expedition.
3 Skelton knew Scott from HMS *Majestic*, where Scott served under Egerton (a veteran of Nares's Arctic expedition). Shackleton learned of the *Discovery* expedition when he met the son of expedition sponsor Longstaff on a Boer War troopship.
4 Industrialist Llewelyn Longstaff donated £25,000, but the Admiralty's refusal to provide ships strained the expedition's finances considerably.
5 The rolling was a consequence of opting for a retractable rudder.
6 Beaumont, another veteran of Nares's Arctic expedition, seconded Tom Crean and provided a last-minute replacement for seaman Bonner, who died after falling off the top mast as *Discovery* left Lyttelton.
7 It is unclear whether they passed Borchgrevink's record.
8 The hut was never intended for long-term occupation.
9 Scott wondered if the dried stockfish Nansen recommended was affected by passing through the tropics.

10 Sledging rations of pre-cooked seal meat lacked anti-scorbutic vitamin C.
11 Scott's report describes Shackleton's service as 'highly satisfactory' but his 'constitution ha[d] proved unequal to the rigours of a polar climate'. There seem to have been no other disagreements, but while Shackleton was naturally reluctant to leave, Scott needed all officers to be fit enough to lead sledge parties.

26: Edward Wilson's Portable Paintbox

Specific sources: Wilson, David M. & C.J.; Scott, Robert F. (1905); Cherry-Garrard et al. (*South Polar Times*).

1 Scott, *The Voyage of the Discovery*, Vol. 2, p. 53.
2 Wilson painstakingly replicated about eighty paintings for very modest remuneration.

27: A Postcard of Three Scottish Scientists

Specific sources: Speak (2003); Pirie et al.; Williams & Dudeney; Shetland Museum and Archives (Carol Christiansen); George Waterston Memorial Centre & Museum, Fair Isle (Anne Sinclair).

1 The aegis of Bruce's expedition is complex but, in summary: after Markham failed to raise sufficient funds for two ships, he decided to go ahead with one ship (against advice from John Murray). When Bruce raised funds for a second ship, Markham suggested Bruce surrender the funds to Markham. As Bruce's funds were from Scottish supporters and his main interest was science, he and Murray organised an expedition with a similar ethos to the *Challenger* expedition (on which Murray and other Scottish scientists served), with a distinctive Scottish identity.
2 Bruce's name is sometimes given as 'Speirs Bruce'; other expedition members sometimes use middle names, but I follow Speak (2003) in using 'Bruce', 'Pirie' and 'Rudmose Brown'.
3 The brothers, who regularly sailed to the Shetlands, bought large quantities of Fair Isle and Shetland knitwear.
4 Bruce was born in London to a Scottish father and Welsh mother; he was schooled in England but studied at Edinburgh University and considered himself Scottish.
5 The Orkneys are at approximately 60°N.
6 Bruce's mentor, Robert Omond, was the director of Edinburgh's observatory.
7 Pirie uses the Scottish word 'dreich' (in quote marks) in the expedition report.
8 Bruce later co-founded the Scottish Ski Club.
9 Speak, p. 87, quotes Bruce on the effect of freezing.
10 Fair Isle winter gear is specifically praised in the expedition report.
11 A suggested reason for the refusal was that Bruce had no mandate to claim Laurie Island on which Omond House stood.
12 Film clips from the *Scotia* expedition (on the National Library of Scotland's website) are the earliest known films of Antarctica; Shackleton's films were the first distributed commercially.
13 Wilson's copy of the book was later transferred to Gloucestershire Archives (B153/50753GS).

14 Although Markham may have pointedly not awarded Polar Medals to *Scotia* expedition members, members of the privately funded Jackson-Harmsworth expedition did not receive Arctic Medals.

28: ARA *Uruguay*

Note: This is a brief tribute to the ship and men involved in one of the most remarkable, but relatively little known, search-and-rescue efforts in polar history.
Specific sources: Markham/Irízar (paper); Pirie et al., ch. XVII; Rosove et al.
1 ARA (*Armada de la República Argentina*) equates to Britain's HMS.
2 The Swedish authorities apparently declined Argentinian offers of assistance on the grounds that the Argentinians lacked polar experience (*Lloyds Weekly Newspaper*, 28 June 1903).
3 Shackleton also worked on *Terra Nova*'s relief of *Discovery*.
4 Sobral served under Irízar on British-built gunship ASA *Patria* (*St James's Gazette*, 3 July 1903).
5 Charcot (on *Français*) met *Frithjhof* in Madeira in September 1903 (*Lloyds List*, 22 September 1903); de Gerlache initially accompanied Charcot, but left *Français* before she reached Buenos Aires.
6 The northern end of the Antarctic peninsula was described both as Trinity Land (Bransfield) or Louis Philippe Land (d'Urville).
7 *Frithjhof* left on 17 January (*Liverpool Journal of Commerce*, 19 February 1904).

Part VI: Striving for Polar Firsts

29: Amundsen's Dip Circle

Specific sources: Amundsen (1908 and 1925); Bown; Bomann-Larsen; Riffenburgh (2010); Jacobsen; Fram Museum.
1 Amundsen (1908) uses *Gjöa*; the Fram Museum and others use *Gjøa*.
2 Donors (including Britain's Royal Geographical Society) are listed in Amundsen (1908), addendum.
3 Amundsen (1908), Chapter II.
4 See also Russell Potter's 'Visions of the North' post 'Lost and Found: The Beechey Island Anvil Block'.
5 Amundsen's magnetic readings were planned to coordinate with those by Scott and others in Antarctica but did not always do so due to mistaken time differences (Bown, Chapter 4).
6 Per Stenton, D.R. (paper), American Charles Hall found the site in 1869.

30: Shackleton's Sledging Compass

Specific sources: Shackleton (1909); Riffenburgh (2004); Daly; Shackleton biographies (Fisher, Huntford (1985), Mill, Smith (2014)); South Georgia Museum; Wilson, D.M. (2009)
1 Shackleton probably met Beardmore through the RGS. Some suggest Beardmore, like Shackleton, attended Dulwich College, but no record of his name has been found in college records (although there is a nearby Dulwich Prep School).

2 Jackson and Armitage suggested the use of ponies based on their Arctic experience but Nansen advised against their use.
3 Dead reckoning involves steering by compass and calculating elapsed mileage and would not normally be regarded as the basis for a claim by the RGS or similar organisations; Shackleton, who felt it vital to conserve the men's strength, knew he was uncontestably ahead of Scott's *Discovery* record.
4 Forced March tablets contained coca leaf (a known appetite suppressant used to make cocaine) and kola nut (a source of caffeine); cocaine was not illegal at the time and coca leaf (as evidenced by the name) was later used in Coca-Cola.

31: Deception Island
Specific sources: Charcot (1910, 1911, 1906/2004); Quilty; public domain information on Deception Island and the Andresens.
1 Pendulum Bay is named for Henry Foster's 1820s pendulum-based magnetic readings.
2 See 'Quite the Guernseyman: Jean-Baptiste Charcot', blog, 6 March 2019 (website of Prieulx Library, Guernsey).

32. Matthew Henson's Fur Suit
Specific sources: Henson (both editions); Peary; Henderson; Robinson; Weems; Evans; Berkshire Museum, Pittsfield, Mass. USA.
Author's note: Accounts of Peary and Cook's achievements vary widely based on the time of writing and the authors' existing viewpoints (see Stam's 2008 introduction to the 2016 edition of Henson's *A Journey for the Ages*).
1 The survey in Nicaragua was a prelude to the construction of the Panama Canal.
2 Cook objected to Peary's attempts to control publication of his scientific findings and later travelled to Greenland with others.
3 Peary recorded 86° 6'N but did not formally claim a new Farthest North.
4 See Henderson, *True North* (Chapter 16) for more detail.
5 Peary's account and Henson's biography both give the first sightings as 89° 57'N.
6 Peary included the photograph of Henson and the four Inuit in his expedition report.
7 Henson suggests in his 1912 memoir that Peary set a surprisingly fast pace on the return journey; the speed, as mentioned in Robinson, *Dark Companion*, was one of the reasons Peary's claim was challenged.
8 National Geographic Society (founded 1888, now also known as National Geographic) sponsored Peary's expedition and immediately accepted his claim; the less-entrepreneurial American Geographical Society (founded 1851) did not validate his claim.
9 Inuit expedition members were not awarded medals; it is unclear whether this related to citizenship requirements or to their being largely overlooked in Peary's records.

Part VII: Southward Ho!
1 *Field*, 22 January 1910.

33. Ponting's Kinematograph
Specific sources: Strathie (all books); Ponting; Arnold (both books); Attwood (paper); Pritchard; BNA newspaper reports re. expedition departure.
1. The film magazine issue with Borchgrevink's 1898 *Southern Cross* expedition camera had been resolved.
2. Prestwich repurchased the kinematograph following the expedition and gave it to the Science Museum (Bradford).
3. Meares joined Ponting on a photographic tour of India, Myanmar and Sri Lanka; he had previously worked and driven dog sledges on the Kamchatka Peninsula.
4. Photography was an additional duty for Bernacchi, Skelton and other photographers.
5. Films could be heat damaged when passing through the tropics.
6. Ponting's illustrated Japanese memoir, *In Lotus Land: Japan*, was published the day *Terra Nova* sailed; Ponting was also exhibiting photographs at the Japan–British Exhibition at White City.
7. Ponting was not party to the Gaumont contract but was credited as filmmaker ('H.G. Ponting, FRGS', elected in 1905).
8. It appears Ponting's Antarctic stereoview images may not have been published.
9. Scott named Cape Evans for Edward 'Teddy' Evans, formerly *Morning*'s navigator, who planned his own Antarctic expedition but (at Clements Markham's instigation) joined forces with Scott and was appointed both Scott's naval No. 2 and *Terra Nova*'s captain.

34. A Samurai Sword
Specific sources: Shirase Antarctic Expedition Supporters' Association (SAESA); Turney.
1. SAESA, p. 383, note 4.
2. SAESA, pp. 22–23; it is unclear whether Shirase, who had previous Arctic experience, initially planned to try for the North Pole but changed plans following Cook and Peary's claims.
3. Ōkuma (whom Ponting photographed during the Russo-Japanese War) strongly supported Japan's interaction with the wider world.
4. Victor Campbell's now Northern Party landed at Cape Adare a few weeks previously.
5. Press reports reflect then-held concerns regarding Japanese ambitions following their recent victory over Russia; SAESA editors suggest linguistic difficulties, Shirase's shortage of funds and his and his men's feelings of failure also contributed to their initially feeling unwelcome.
6. After Mawson decided against joining Scott's *Terra Nova* expedition, he briefly collaborated on fundraising with Shackleton, but found the latter an unreliable partner.
7. The 1902 Anglo-Japanese Alliance remained in force. Turney (pp. 157–58) quotes the full letter.
8. Yamato means 'great harmony', an ancient local name for Japan.
9. As Pathé retained the bulk of film revenues and profits, Shirase, like many of his fellow explorers, remained in debt.

35. Mawson's Anemograph
Specific sources: Riffenburgh (2010 and 2011); Turney; Bradford.
1. Katabatic winds blow from high to low land. Cape Denison gusts regularly exceeded 150mph and monthly winter averages were around 55mph; the *Guinness Book of Records* (2023) still includes the Cape Denison record wind speeds.
2. Hurley, who regularly took risks to obtain a dramatic photograph, was trained to use a kinematograph by Fred Gent, then at Gaumont's Sydney office.
3. Most sources suggest Mertz suffered vitamin A poisoning from eating dog livers.
4. Mawson expected Wild's party to be 300–500 miles west, but conditions prevented planned landings.

36. Joseph Kinsey's Visitors' Book, April 1912
Specific sources: Strathie (all books); MacInnes (both books); Fletcher; Lummis (paper); PapersPast website (New Zealand); Harry Pennell's journals (Canterbury Museum, MS107, MS433).
1. Pennell became captain of *Terra Nova* after Teddy Evans and first officer Victor Campbell landed.
2. Shirase had returned; Mawson's *Aurora* was back, with Mawson overwintering again; Filchner was possibly iced in.
3. See also Strathie (2015) and RGS's 'HLP'/Pennell file. In summary: Evans asked Scott in late 1911 if he could leave the landing party and rejoin his wife in Christchurch. It is unclear whether Evans intended to leave the expedition completely, but Scott gave Evans permission to return south on 4 January 1912. Evans was expected to captain *Terra Nova* back to New Zealand, leaving Pennell free to join the shore party, but after he developed scurvy, Pennell had to remain with the ship. Crean and Lashly received Albert Medals for saving Evans's life.
4. Amundsen considered returning to Lyttelton but switched to Australia following criticism from Clements Markham and others, who regarded his switch from North to South Poles as a breach of polar etiquette.
5. Evans's dislike of and refusal to eat seal meat was probably the main contributing factor to his developing it while others did not.
6. Fletcher, p. 127; MacInnes (2022), pp. 251–52.
7. Amundsen's signature is on an earlier page of the visitors' book, along with those of *Terra Nova* expedition members.

37. 'Three Polar Stars' Photograph, January 1913
Specific sources: Fletcher, MacInnes (both books); Young; Kennet/Scott; Jan-Emil Kristofferson/Strathie research on Martin's supper; Shackleton biographies (Fisher, Huntford (1985), Mill, Smith (2014)); Bown; Bomann-Larsen; Geographical Society of Philadelphia website.
1. The US edition of Shackleton's *Nimrod* narrative was published by J.B. Lippincott of Philadelphia; it is not known whether members of the Lippincott family were members of the city's geographical society.
2. Telegram details from Carnegie Hall's online event records.

3   Peary already held the RGS 1898 Patron's Medal (Greenland expeditions), but published information and RGS records suggest the Peary Medal was and remains controversial. Kathleen Scott attended Amundsen's RGS lecture in London but not in her capacity as Scott's wife, so had not been formally introduced to him.
4   Peary hoped MacMillan would return from Crocker Land with geological samples and charts which would disprove Cook's claim, but MacMillan found no evidence that Crocker Land existed.
5   The two had a quarrel long before the final departure for the pole, following which Amundsen excluded Johansen from the final South Pole party and excluded him from any further decision-making during the expedition.

38. Henry 'Birdie' Bowers's Sledge Flag
Specific sources: Strathie (2010 and 2015, where SPRI/MS1505 and other references are cited in full); MacInnes (both books); Seaver (1938); SPRI/Polar Museum blog 'Men Who Sew, Part 3: Birdie Bowers' (10 December 2014/2023); Tomlinson (paper).
1   Bowers, as a young schoolboy, wrote to an inhabitant of Wilkes Land, expressing hopes of visiting him. While studying at and visiting HMS *Worcester*, he attended lectures by Clements Markham and Albert Armitage (of *Discovery*).
2   Bowers to family, 25 September 1910 (see SPRI/MS1505).
3   Bowers to family, 28 November 1910; Bowers suggested the crest, sometimes used by his father, may have related to ancestors who were archers (bow-ers).
4   The tent contained a letter addressed to Scott, asking him to bring back a letter marking Amundsen's achievement in case anything happened to Amundsen on his return journey.
5   Neither Oates nor Edgar Evans had sledge flags; Bowers's and other sledge flag designs were painted by Wilson and published in 'South Polar Times' (Cherry-Garrard, et al.)
6   Oriana Wilson to Emily Bowers, 22 May 1911.
7   Seaver had previously written a biography of Wilson. The sledge flag's plaque does not indicate whether Bowers's sister presented it in person, but Oriana Wilson and Scott's sister both gave Scott Memorial lectures at the school.

Part VIII: 'White Warfare' and Testing Times
1   Donald MacMillan's Crocker Land expedition failed to underpin Peary's claim.

39. An Expedition Prospectus
Specific sources: Prospectus (as described in chapter text); Shackleton biographies (Fisher, Huntford (1985), Mill, Smith (2014)); Shackleton/Riffenburgh (*South*); Strathie (2015); BNA newspaper reports.
1   It is unclear who Shackleton meant by the 'generous friend', who he said made an early donation; Mill and Huntford (1985) suggest candidates, but Huntford also suggests it could be a pseudonym for the British government.
2   The first print run names John K. Davis as ship's captain, but he declined to participate in the expedition.
3   Shackleton purchased *Endurance/Polaris* from the Norwegian shipyard; Mawson sold *Aurora* to clear expedition expenses.

4 The amount of Janet Stancomb-Wills's donation remains undisclosed.
5 Caird to Shackleton, 17 June 1914, SPRI/1537/2/30/22:D (online), James Caird Society website; Shackleton mentions £10,000 being required in both prospective printings.
6 Shackleton purchased a Falklands sealing concession before leaving England. Had he stopped at the Falklands, he could have been held there during the Battle of Coronel (off Chile, November, British defeat) and/or the Battle of the Falkland Islands (December, British victory).
7 Some of the expedition narratives were in the ship's library.

40. A Statue of Cheltenham's Local Hero
Specific sources: Strathie (2015, 2021); MacInnes (both books); Fletcher; Kennet/Scott; Young; The Wilson Museum & Art Gallery archives.
1 It is unclear whether the Scotts were aware that the awarding of the medal to Peary (who had been awarded the RGS's 1898 Patron's Medal for previous expeditions) was the subject of much internal debate within the RGS.
2 The wallet, now at SPRI (ref. N: 1045), was loaned for a 2023–24 exhibition on Wilson in Cheltenham.

41. A Rock from Elephant Island
Specific sources: Smith (2004/7, which includes Wordie's *Endurance* log); Sharpe (paper); Dudeney (paper).
1 Gregory was familiar with the *Discovery* expedition plans. Sledge flag: Smith, p. 34, Kennings, who made freemasonry and naval banners, could have been suggested by sledge flag enthusiast Clements Markham.
2 When they arrived, seals and penguins were heading north for the winter.
3 See Dudeney et al. for more on Shackleton's own and official British rescue attempts. In short, Shackleton would have lost revenue from newspapers and other supporters if he had agreed to the terms under which the government were prepared to rescue him.
4 The party included Wild's younger brother Ernest and (as Wordie may have known) another Gregory protégé, geologist Alexander Stevens.

Part IX: The Age of Aviation
1 Norway remained neutral but Gran had, during the *Terra Nova* expedition, promised Bowers and Oates he would fight for Britain during an increasingly likely war with Germany (see Strathie, 2015, Prologue).

42. An Avro Antarctic Baby
Specific sources: Wild & Macklin; Chojecki; Noake; Marr; Jackson, A.J./R.T., *Avro Aircraft since 1908* (London: Putnam/Conway, 1990/1965); BNA (on Carr's racing).
1 Following the war, Wilkins served on Vilhjalmur Stefansson's 1913–16 Arctic expedition and J.L. Cope's truncated 1920 Antarctic expedition.
2 BNA articles, September 1922; report to Rowett, August 1922.

43. A Tribute to Shackleton from a 'Fan'

Specific sources: Service, Robert, *Songs of a Sourdough* (original UK edition, London: Fisher Unwin, 1907); Daly; Mill; Mayer; Kelly, Ethel K., *Twelve Milestones* (London: Brentano's, 1929), Chapter/Milestone IV.

Service's words are quoted with kind permission of Mme Anne Longépé and Charlotte Service-Longépé.

1 The original, 'The Heart of the Sourdough', reads: 'By all that the battle means and makes I claim that land for mine own/Yet the Wild must win, and a day will come when I shall be overthrown.' It is unclear whether Shackleton or Russell Gregg was the source of the variation.
2 Service was born in Preston and lived in Scotland before immigrating to Canada.
3 Shackleton to Mill (Daly, Letter 126).
4 It is unclear whether Russell Gregg attended the service.
5 See 'Fancy Dress at SPRI', SPRI blog, 8 November 2010. In 1923, Cambridge Festival Student Creatives (including Rebekah King) undertook a project on the costume and ball, with assistance from members of the Chelsea Arts Club.
6 Mill, p. 125.
7 Mill, pp. 166–67.
8 Mill, p. 174.

44. 'Uranienborg': An Explorer's Refuge

Specific sources: Bomann-Larsen; Kafarowski; MiA/Uranienborg website and information provided by Anders Bache.

1 Amundsen technically lost possession of 'Uranienborg' after being declared bankrupt, but his friends and supporters Herman Gade and 'Don Pedro' Christophersen repurchased it so he could continue living there.
2 American Richard Byrd claimed to have reached the North Pole a few days before Amundsen, but Amundsen and others had doubts about Byrd's recorded data. Amundsen did not air his doubts in public as he was still the first man to both poles.
3 Stokes travelled with Peary (Greenland) and Nordenskjöld (Antarctica).
4 Research by the author and Anders Bache suggest the reproduction may have been given to Amundsen by a British friend; this may have been Teddy Evans of the *Terra Nova* expedition, who met Amundsen following Scott's expedition, married into a well-known Norwegian family, lived part-time in Norway and chaired at least one of Amundsen's London lectures in the 1920s.
5 Turney, 1912, p. 173.
6 In summary, Amundsen visited Cook, who was in jail for suspected fraud. Comments Amundsen made to reporters following the visit suggested to NGS's officials (Peary's greatest supporters and expedition sponsors) that Amundsen now supported Cook's claims. Amundsen maintained he had been misquoted and, in his memoir, supported Peary's claim. On a personal basis, however, he evidently felt sorry for Cook and was touched when Cook gave him a hand-embroidered table runner he had made for Amundsen (still on display at 'Uranienborg').

7   No formal record has been found which details what Curzon said; at best, he appears to have been tactless or made an ill-judged attempt at humour, which Amundsen took as a slight.
8   Huntford (2000, p.664) and Drivenes et al. (p.276) suggest Amundsen received radium treatment (possibly for a tumour on his leg) around this time.

45. Mawson's Gipsy Moth
Specific sources: Mawson (paper); McGregor; Museums Victoria collections information.
1   The Hudson's Bay Company purchased Scott's *Discovery*, then resold it to the British government, which loaned it, including to Mawson, for polar expeditions.
2   As Mawson had hoped, the claim was later transferred to Australia.
3   Norway claimed Queen Maud Land.
4   The South Magnetic Pole continued to move north.

46. A Young Explorer's Special Medal
Specific sources: Scott, J.M.; Chapman; Hodges, John R., *Dumbleton Hall* (2015, private publication).
1   The medals listing reflects the order of their awarding/presentation.
2   Robert Scott's son Peter undertook several roof climbs with Watkins. Peter Scott's step-uncle, Geoffrey Winthrop Young, wrote an 1899 guide to Cambridge roof climbing.
3   Wegener (1880–1930) worked on continental drift theories, the forerunner of modern plate tectonics; material relating to his work is held by Uumannaq Museum, Greenland.

Part X: Learning From the Past and Looking to the Future

47. A 'Polar Centre': The Scott Polar Research Institute Building
Specific sources: Debenham (paper); Speak (2008); Kennet/Scott; Bernacchi et al.; Strathie (2021); Walker.
1   The phrase may originate from an epitaph coined by Tennyson's friend and fellow poet William Landor (1775–1864); '*Patriae quaesivit gloriam videt Dei*' appears on several headstones in Great War cemeteries.
2   The British Empire Film Institute agreed to purchase Ponting's master tapes of expedition films 'for the nation' in 1929; they received financial assistance from the Hudson's Bay Company (of which Ponting's friend Charles V. Sale was governor) and Alfred Bossom (a friend of Ponting) but did not pay Ponting the full agreed amount.
3   MacDonald Gill (Eric Gill's younger brother) previously worked with Baker and was famous for decorative maps, including of the London Underground and the British Empire. Betty Creswick, Debenham's assistant, provided initial watercolour sketches of specific ships as required.
4   It is unclear whether anyone represented Oates or Edgar Evans.
5   A bronze plaque of Watkins (financed by an anonymous donor) was installed in SPRI's museum around 1935.

48. The *Erebus* Bell
Specific sources: Potter (2016); Geiger & Mitchell; Parks Canada online information; 'Visions of the North' blog (Potter); 'If any living Inuk knew', interview with Louis Kamookak (*Up Here*, December 2014).
1. At the time of writing, *Terror*'s bell has been located but has not been raised.
2. If daguerreotypes were taken and stored in sealed, watertight boxes some traces of the images on them might remain.

49. An Expedition Hut
Specific sources: Strathie (all books); Antarctic Heritage Trust New Zealand website; BNA newspaper articles, including *Illustrated London News*, 16 July 1958.
1. Strathie, A. (2015), p. 152.
2. The RGS has Agfa flashlamp refill pack (RGS 700926) and copies of photographs with supporting documents (RGS PR/052791, A and B).
3. Clergymen were exempt from First World War military duties.

50. A Well-Travelled Crow's Nest
Specific sources: Chojecki; Marr; South Georgia Museum website (including about *Quest* crow's nest tour); BNA newspaper reports; Simon Louagie, Talbot House, Poperinghe, Belgium.
1. *Quest*'s previous name was *Foca* (Norwegian for seal) but she remained *Quest* until she sank off Labrador in 1962.
2. *Daily News*, 25 January 1923.
3. *Western Morning News*, 4 January 1930.
4. *Derby Daily Telegraph*, 15 April 1930.
5. See Clayton's letter to his mother, 4 January 1917 (information from Simon Louagie, Talbot House); Andersen wrote to Ponting saying how much soldiers enjoyed the films (Strathie, 2021, pp. 143–44 and related endnote 48).
6. Hurley took the now-famous photograph (McGregor, p. 152) on 3 October 1917; it is not known if they visited Talbot House, but Clayton may have met Wilkins in 1922 and made the wartime connection.
7. See pp. 250–1 Bullard, Reader, *Letters from Tehran* (London: I.B. Tauris, 1991) re. Clayton and Hurley's encounter.

Bibliography

Due to the timespan and range of objects, people, expeditions and themes encompassed in *A History of Polar Exploration in 50 Objects*, this bibliography is extensive. The majority of books and papers are those read or consulted (including via online providers) in relation to the main text of this book; these are also cited, by author name, under 'specific sources' headings in the Notes section. Other books and papers provided valuable background information or are recommended reading for those interested in the history of polar exploration. More detailed bibliographies relating to Robert Scott's *Terra Nova* expedition and its members appear in my previous books.

Airriess, Sarah, *The Worst Journey in the World, Vol. 1: Making our Easting Down* (London: Indie Novella, 2022).
Amundsen, Roald, *My Life as an Explorer* (London: William Heinemann Ltd, 1927).
— *My Polar Flight* (London: Hutchinson & Co., 1925).
— *The Northwest Passage* (London: Constable, 1908).
— *The South Pole* (London: C. Hurst, 1976, originally published by John Murray, 1912).
Amundsen, Roald, et al. (Le Piez, Christine, & Hugo Declair, translated by Erik Dupont), *Roald Amundsen's Belgica Diary* (Norwich: Erskine Press; Huntington: Bluntisham Books, 1999).
Arnold, H.J.P., *Herbert Ponting: Another World* (London: Sidgwick and Jackson, 1975).
— *Photographer of the World* (London: Hutchinson & Co. (Publishers) Ltd, 1969).
Batchen, Geoffrey, *Apparitions: Photography and Dissemination* (Sydney/Prague: Power Publications/AMU Press, 2018).
Battersby, William, *James Fitzjames* [Franklin expedition] (Cheltenham: The History Press, 2010, 2023).
Beaglehole, J.C., *The Life of Captain Cook* (London: A. & C. Black/Hakluyt Society, 1974)
Beattie, Owen, & Geiger, John, *Frozen in Time: The Fate of the Franklin Expedition* (London: Bloomsbury, 1987, 2004).
Bernacchi, Louis, *A Very Gallant Gentleman* [Lawrence Oates] (London: Thornton Butterworth, 1933).
Bernacchi, Louis, et al. (including Mill, Debenham, Wordie, Rudmose Brown, Watkins), *The Polar Book* (London: E. Allom & Co. Ltd, 1930).

Bomann-Larsen, Tor (translated by Ingrid Christophersen), *Roald Amundsen* (Stroud: Sutton Publishing/The History Press, 2006).

Bound, Mensun, *The Ship Beneath the Ice: The Discovery of Shackleton's Endurance* (London: Macmillan, 2022).

Bown, Stephen, *The Last Viking: The Life of Roald Amundsen* (London: Aurum Press Ltd, 2012).

Bradford, Karyn Maguire, *The Crevasse* [Mawson *Aurora* expedition] (Norwich: Erskine Press, 2015).

Browne, W.H., et al., *Ten Coloured Views Taken During the Arctic Expedition of Her Majesty's Ships 'Enterprise' and 'Investigator', Under the Command of Captain Sir James C. Ross* (London: Ackermann and Co., 1850).

Brunton, Eileen V., *The Challenger Expedition, 1872–6: A Visual Index*, 2nd Edition (London: Natural History Museum, 2004, online).

Bulkeley, Rip, *The Historiography of the First Russian Expedition, 1881–20* (London: Palgrave Macmillan, 2021).

Burford, Robert, et al., *Description of Summer and Winter Views of the Polar Regions* (London: Golbourn, 1850).

Butler, Angie, *The Quest for Frank Wild* (Warwick: Jackleberry Press, 2011).

Campbell, Victor (edited by H.G.R. King), *The Wicked Mate: The Antarctic Diary of Victor Campbell* (Norwich: Erskine Press; Huntington: Bluntisham Books, 1988).

Capelotti, P.J., *Shipwreck at Cape Flora: The Expeditions of Leigh Smith, England's Forgotten Arctic Explorer* (Calgary, Canada: University of Calgary Press, 2013, available online).

Captain Cook Memorial Museum (invited authors), *Ice! Exploring the Far South: 250th Anniversary of the first crossing of the Antarctic Circle* (Whitby: Captain Cook Memorial Museum, 2023).

Chapman, F. Spencer, *Watkins' Last Expedition* (London: Chatto & Windus/Heinemann, 1934/1953).

Charcot, Jean-Baptiste, *Dans la Mer du Groenland: Croisières du 'Pourquoi Pas?'* (Bruges: Librairie de l'Oeuvre Saint-Charles, 1935).

— *Le Pourquoi-Pas? dans l'Antarctique, 1908–1910* (Paris: Flammarion, 1910).

— *The Voyage of the 'Why Not?'* (London: Hodder and Stoughton, c.1911).

— (translated by A.E. Billinghurst) *Towards the South Pole aboard the Français* (Norwich: Erskine Press; Huntington: Bluntisham Books, 2004; orig. *Le Français au Pôle Sud*, 1906)

Cherry-Garrard, Apsley, *The Worst Journey in the World* (London: Vintage Classics, 2010 and 1922).

Cherry-Garrard, Apsley, et al., *The South Polar Times* (London: The Folio Society, 2018 edition; Smith, Elder, originally published 1901–13).

Chojecki, Jan, *The Quest Chronicle: The Story of the Shackleton-Rowett Expedition* (Cheltenham: Goldcrest Books, 2022).

Claydon, Annalise Jacobs, *Arctic Circles and Imperial Knowledge* [Franklin expedition and family] (London: Bloomsbury Academic, 2024).

Conan Doyle, Arthur, *Dangerous Work: Diary of an Arctic Adventure* (London: The British Library Publishing, 2012).

Cook, James (edited by Philip Edwards), *The Journals* (London: Penguin Books, 1999).

Crane, David, *Scott of the Antarctic* (London: HarperCollins, 2005).
Crawford, Janet, *That First Antarctic Winter: The Story of the Southern Cross Expedition of 1898–1900 as Told in the Diaries of Louis Charles Bernacchi* (Christchurch, NZ: South Latitude Research Ltd/Peter J. Skellerup, 1998).
Cyriax, Richard J., *Sir John Franklin's Last Arctic Expedition* (London: Methuen, 1936, reprinted Plaistow: The Arctic Press, 1997).
Daly, Regina (ed.), *The Shackleton Letters: Behind the Scenes of the Nimrod Expedition* (Norwich: Erskine Press, 2009).
David, Robert G., *The Arctic in the British Imagination* (Manchester: Manchester University Press, 2000).
De Gerlache de Gomery, Adrien (translated by Maurice Raraty), *Voyage of the Belgica: Fifteen Months in the Antarctic* (Norwich: Erskine Press; Huntington: Bluntisham Books, 1998).
Debenham Frank, *Antarctica* (London: Herbert Jenkins, 1959).
— (edited by June Back), *The Quiet Land* (Norwich: Erskine Press; Huntington: Bluntisham Books, 1992).
Dodge, Ernest S., *The Polar Rosses* (London: Faber & Faber, 1973).
Douglas, M., *The White North* [Nansen, et al.] (London: Thomas Nelson & Sons, 1899).
Drivenes, H. Dag, et al., *Into the Ice: The History of Norway and the Polar Regions* (Copenhagen: Gyldenal Norsk Forlag, 2006).
Evans, Richard, *The Explorer and the Journalist: Frederick Cook, Philip Gibbs and the Scandal that Shook the World* (Cheltenham: The History Press, 2023).
Fisher, Margery & James, *Shackleton* (London: James Barrie Books, Ltd, 1957).
Fleming, Fergus, *Barrow's Boys* (London: Granta Books, 1998).
Fletcher, Anne, *Widows of the Ice: The Women That Scott's Antarctic Expedition Left Behind* (Stroud: Amberley Publishing, 2022).
Forgan, Sophie, et al., *Smoking Coasts and Ice-Bound Seas: Cook's Voyage to the Arctic* (Whitby: Captain Cook Memorial Museum, 2017).
Franklin, Jane (edited by W.F. Rawnsley), *The Life, Diaries and Correspondence of Jane Lady Franklin* (Cambridge: Cambridge University Press, 2014, orig. 1923).
Franklin, John, *Narrative of a Journey to the Shores of the Polar Seas, 1819–22* (London: J.M. Dent & British Museum, 1823, 1970).
Fretwell, Peter, *Antarctic Atlas* (London: Particular Books, Penguin/Random House, 2020).
Fuchs, Vivian, & Hillary, Edmund, *The Crossing of Antarctica* (London: Cassell & Company Ltd, 1958).
Geiger, John, & Mitchell, Alanna, *Franklin's Lost Ship* (Toronto, Canada: HarperCollins, 2015).
Goodsir, Robert, *Arctic Voyage in Search of Franklin* (London: J. Van Voorst, 1850; reprint, Plaistow & Sutton Coldfield: The Arctic Press, 1996).
Hackney, Roan, & Budgen, Marcus, *200 Years of Polar Exploration, 1819–2019* (London: Spink, 2019).
Hamilton, James C., *Captain James Cook and the Search for Antarctica* (Barnsley/Philadelphia, Pen & Sword Books, 2020).
Hayes, J. Gordon, *The Conquest of the North Pole* (London: Thornton Butterworth Ltd, 1934).

— *The Conquest of the South Pole* (London: Thornton Butterworth Ltd, 1932).
Hempleman-Adams, David, et al., *The Heart of the Great Alone* (London: Royal Collection Enterprises Ltd, 2009, 2011).
Henderson, Bruce, *True North: Peary, Cook and the Race to the Pole* (New York/London: W.W. Norton & Company, 2005).
Henson, Matthew, *A Negro Explorer at the North Pole* (California: Mint Editions, 2012 reprint, orig. 1912).
Henson, Matthew, et al., *A Journey for the Ages*, Explorer Club Classic (New York: Skyhorse Publishing, 2016).
Hough, Richard, *Captain James Cook: A Biography* (London: Hodder & Stoughton Ltd, 1994).
Hoyle, Gwyneth, *Flowers in the Snow: The Life of Isobel Wylie Hutchison* (Nebraska: University of Nebraska Press, 2011).
Huntford, Roland, *Nansen* (London: Abacus; Gerald Duckworth & Co., 2001, 1997).
— *Scott and Amundsen: The Last Place on Earth* (London: Abacus, 2000).
— *Shackleton* (London: Hodder and Stoughton, 1985).
Hurley, Frank, *The Enduring Eye* (London: RGS with IBG, 2015).
Hurley, Frank (edited by Robert Dixon & Christopher Lee), *The Diaries of Frank Hurley, 1912–41* (London: Anthem Press, 2011).
Hutchinson, Gillian, *Sir John Franklin's Erebus and Terror Expedition: Lost and Found* (London: Bloomsbury and Royal Museums Greenwich, 2017).
Jackson, Frederick G., *A Thousand Days in the Arctic* (London; New York: Harper & Brothers, 1899).
Jacobsen, Anne Christine, & Huntford, Roland (eds), *The Amundsen Photographs* (London: Hodder & Stoughton, 1987).
James, David, *Scott of the Antarctic: The Film and its Production* (London: Convoy Productions, 1948).
Jones, A.G.E., *Antarctica Observed: Who Discovered the Antarctic Continent* (Whitby, Yorkshire: Caedmon of Whitby, 1982).
Jones, Erika, *The Challenger Expedition: Exploring the Ocean's Depths* (London: Royal Museums Greenwich, 2022).
Jones, Max, *The Last Great Quest* (Oxford: Oxford University Press, 2003).
Kafarowski, Joanna, *The Polar Adventures of a Rich American Dame: A Life of Louise Arner Boyd* (Toronto: Dundurn, 2017).
Kennet, Kathleen (formerly Scott), *Self-Portrait of an Artist* (London: John Murray, 1949).
Knight, John, *The Magnetism of Antarctica* (Caithness, Scotland: Whittles Publishing, 2023).
Lamb, G.F., *Franklin: Happy Voyager* (London: Ernest Benn Ltd, 1956).
Lambert, Andrew, *Franklin: Tragic Hero of Polar Navigation* (London: Faber and Faber Ltd, 2009).
Le Brun, Dominique, *Charcot* (Paris: Tallandier, 2021).
Lecointe, Georges (translated by Cynthia Kaiser), *In the Land of Penguins* [*Belgica* expedition] (Norwich: Erskine Press, 2020/1904).
Lefroy, John Henry (edited by George F.G. Stanley), *In Search of the Magnetic North* (Toronto: Macmillan, 1955).

Lewis-Jones, Huw, *Imagining the Arctic: Heroism, Spectacle and Polar Exploration* (London: I.B. Tauris & Co. Ltd, 2017).
Limb, Sue, & Cordingley, Patrick, *Captain Oates: Soldier and Explorer* (Barnsley: Pen & Sword Books, 1982, rev. 2009).
McClintock, F.L., *The Discovery of the Fate of Sir John Franklin and His Companions* (London: John Murray, 1859).
McGoogan, Ken, *Fatal Passage* [John Rae] (Toronto/London: HarperCollins/Random House/Bantam, 2001)
— *Lady Franklin's Revenge* (London: Random House/Bantam, 2006).
— *Searching for Franklin* (British Columbia, Canada: Douglas & McIntyre, 2023).
McGregor, Alasdair, *Frank Hurley: A Photographer's Life* (Camberwell, VA, Australia: Viking Penguin, 2004).
MacInnes, Katherine, *Snow Widows* (London: William Collins, 2022).
— *Woman with the Iceberg Eyes: Oriana F. Wilson* (Cheltenham: The History Press, 2019).
Markham, Albert, *Life of Sir John Franklin* (London: George Philip & Son Ltd, 1891).
— *The Great Frozen Sea: A Personal Narrative of the Voyage of the Alert ... 1875–6* (London: Daldy, Isbister, & Co., 1878).
Marr, James, *Into the Frozen South* [Shackleton's *Quest* expedition] (London: Cassell & Co., 1923).
Mayer, Jim, *Shackleton: A Life in Poetry* (Oxford: Signal Books, 2014).
Mill, Hugh R., *The Life of Sir Ernest Shackleton* (London: William Heinemann Ltd, 1924).
Mountfield, David, *A History of Polar Exploration* (London: Hamlyn, 1974).
Nansen, Fridtjof, *Farthest North* (London: George Newnes Ltd, 1898).
Nares, George, *Narrative of a Voyage to the Polar Sea ... 'Alert' and 'Discovery'* (London: Sampson Low & Co., 1878).
Noake, Alan, *Shackleton's Scouts* (Sandwich, Kent: 2021).
O'Dochartaigh, Eavan, *Visual Culture and Arctic Voyages: Personal and Public Art and Literature of the Franklin Search Expeditions* (Cambridge: Cambridge University Press, 2022).
Owen, Roderic, *The Fate of Franklin* (London: Hutchison, 1978).
Palin, Michael, *Erebus: The Story of a Ship* (London: Hutchinson, Penguin, 2018).
Parry, W. Edward, *Journal of the Second Voyage for the Discovery of a North-West Passage from the Atlantic to the Pacific* (New York: Greenwood Press, 1969 [rep.], orig. John Murray, 1924).
Peary, Robert, *The North Pole* (New York: Frederick Stokes, 1910).
Pirie, J.H. Harvey, Rudmose, Brown, et al., *The Voyage of the 'Scotia'* (Edinburgh/London: William Blackwood and Sons, 1906).
Ponting, Herbert G., *The Great White South* (London: Gerald Duckworth & Co., 1921).
Potter, Russell A., *Arctic Spectacles* (Seattle/London: University of Washington Press, 2007).
— *Finding Franklin* (Montreal: McGill-Queen's University Press, 2016).
Potter, Russell A., et al. (eds), *May We Be Spared to Meet on Earth* (Montreal: McGill-Queen's University Press, 2022).
Pritchard, Michael, *A History of Photography in 50 Cameras* (London: Bloomsbury, 2015).

Quilty, Patrick, Meidl, Eva, et al. (eds), *The Dawning of Antarctica: Through Exploration to Occupation* (Tasmania: Dr Eva Meidl, 2021–22).

Rae, John (edited and introduction by Ken McGoogan), *The Arctic Journals of John Rae* (Canada: Touchwood Editions, 2012).

Rae, John (introduction by Ken McGoogan), *John Rae's Arctic Correspondence, 1844–55* (Canada: Touchwood Editions, 2014).

Richards, R.L., *Dr John Rae* (Whitby: Caedmon of Whitby, 1985).

Riffenburgh, Beau, *Amundsen* (Cambridge: Scott Polar Research Institute, 2010).

— *Aurora: Douglas Mawson and the Australasian Antarctic Expedition, 1911–14* (Norwich: Erskine Press, 2011).

— (edited by Kløver, Geir O.), *C. A. Larsen: Explorer, Whaler & Family Man* (Oslo: The Fram Museum, 2016).

— *Douglas Mawson* (Cambridge: Scott Polar Research Institute, 2010).

— *Nimrod: Ernest Shackleton and the Extraordinary Story of the 1907–8 British Antarctic Expedition* (London: Bloomsbury, 2004).

— *Terra Nova, Scott's Last Expedition* (Cambridge: Scott Polar Research Institute, 2011).

Robinson, Bradley, *Dark Companion* [Matthew Henson] (London: Hodder & Stoughton, 1948).

Robson, John, *Captain Cook's War & Peace: The Royal Navy Years, 1755–1768* (Barnsley: Seaforth Publishing, 2009).

Rosove, Michael (ed.), Hermelo, Ricardo, et al., *When the Corvette Uruguay was Dismasted: The Return of the Uruguay from the Antarctic in 1903* (Santa Monica CA: Adélie Books, 2004).

Ross, James Clark, *A Voyage of Discovery and Research in the Southern and Antarctic Regions … 1839–43* (Newton Abbot: David & Charles Reprints; orig. London: John Murray, 1847).

Ross, James Clark, et al., *Description of Summer and Winter Views of the Polar Regions* (London: Golbourn, 1850).

Ross, John, *Narrative of a Second Voyage in Search of a North-West Passage … 1833* (London: A.W. Webster, 1835).

— *A Voyage of Discovery in His Majesty's Ships Isabella and Alexander … Probability of a North-West Passage* (London: John Murray, 1819).

Ross, M.J., *Polar Pioneers: John Ross and James Clark Ross* (Montreal: McGill-Queen's University Press, 1994).

Samwell, David, *Captain Cook and Hawaii: A Narrative* (San Francisco; London: David Magee; Francis Edwards, 1957 [reprint from 1786 London edition]).

Sancton, Julian, *Madhouse at the End of the Earth: The Belgica's Journey into the Dark Antarctic Night* (London: W.H. Allen; Penguin, 2021).

Savours, Ann, *The Voyages of the Discovery* (Barnsley: Seaforth Publishing, 2013).

Scoresby, William, *An Account of the Arctic Regions*, Vols 1 and 2 (Trowbridge: David & Charles, 1969 (reprint); Edinburgh/London: Archibald Constable, 1820).

— *A Voyage to the Whale Fishery, 1822* (Whitby: Caedmon of Whitby, 1980 (reprint); London: Archibald Constable, 1823).

— *The Arctic Whaling Journals of William Scoresby the Younger*, Vol. III: 1817, 1818, 1820; (edited by C. Ian Jackson), (London: Routledge; The Hakluyt Society, 2009).

Scott, J.M., *Gino Watkins* (London: Hodder and Stoughton Ltd, 1935).
Scott, Robert F. (edited by Max Jones), *Journals: Captain Scott's Last Expedition* (Oxford: Oxford World's Classics, 2005, 1910–12).
Scott, Robert F., et al., *Scott's Last Expedition*, Vols 1 and 2 (London: Smith, Elder & Co., 1913).
— *The Voyage of the Discovery*, Vols 1 and 2 (London: Smith, Elder & Co., 1905).
Seaver, George, *'Birdie' Bowers of the Antarctic* (London: John Murray, 1938).
— *Edward Wilson of the Antarctic: Naturalist and Friend* (London: John Murray, 1933).
— *Edward Wilson: Nature Lover* (London: John Murray, 1937).
— *Scott of the Antarctic* (London: John Murray, 1940).
Shackleton, Emily, et al., *Rejoice My Heart: The Making of H.R. Mill's 'The Life of Sir Ernest Shackleton'* (Santa Monica, CA: Adélie Books, 2007).
Shackleton, Ernest, & David, T.W. Edgeworth, *The Heart of the Antarctic* (London: William Heinemann, 1909).
Shackleton, Ernest, et al., *The Heart of the Antarctic and South* (Ware, Hertfordshire: Wordsworth Editions, 2007).
Shirase Antarctic Expedition Supporters Association et al., *The Japanese South Polar Expedition, 1910–12* (Norwich: Erskine Press; Huntington: Bluntisham Books, 2001).
Skelton, J.W., & Wilson, D.M., *Discovery Illustrated* (Cheltenham: Reardon Publishing, 2011).
Skelton, Reginald (edited by Judy Skelton), *The Antarctic Journals of Reginald Skelton* (Cheltenham: Reardon Publishing, 2004).
Smith, Michael, *An Unsung Hero: Tom Crean, Antarctic Survivor* (London: Collins Press, 2000, and other editions).
— *Icebound in the Arctic/Captain Francis Crozier: Last Man Standing* (Dublin: O'Brien Press, 2006, 2014, 2021).
— *Polar Crusader: A Life of Sir James Wordie* (Edinburgh: Birlinn Ltd, 2004/2007).
— *Shackleton: By Endurance We Conquer* (London: Oneworld Publications, 2014).
Speak, Peter, *Deb Geographer, Scientist, Antarctic Explorer: A Biography of Frank Debenham OBE* (Guildford: Polar Publishing Ltd/Scott Polar Research Institute, 2008).
— *William Speirs Bruce: Polar Explorer and Scottish Nationalist* (Edinburgh: National Museums of Scotland, 2003).
Spufford, Francis, *I May Be Some Time: Ice and the English Imagination* (London: Faber and Faber, 1996).
Stamp, Tom & Cordelia, *William Scoresby, Arctic Scientist* (Whitby: Caedmon of Whitby, 1975).
Stein, Glenn M., *Discovering the North West Passage ... HMS Investigator and the McClure Expedition* (Jefferson, NC, USA: McFarland & Co., 2015).
Strathie, Anne, *Birdie Bowers: Captain Scott's Marvel* (Stroud: The History Press, 2012).
— *From Ice Floes to Battlefields: Scott's 'Antarctics' in the First World War* (Stroud: The History Press, 2015).
— *Herbert Ponting: Scott's Antarctic Photographer and Pioneer Filmmaker* (Cheltenham: The History Press, 2021).

Tarver, Michael, *The S.S. Terra Nova (1884–1943)* (Brixham: Pendragon, 2006; Cheltenham: The History Press, 2020).
Thomson, C. Wyville, et al., *Report of the Scientific Results of the Voyage of HMS Challenger …*, Vol. 1, Chapter XI (London: Her Majesty's Stationery Office, 1885).
Traill, H.D., *The Life of Sir John Franklin* (London: John Murray, 1896).
Turney, Chris, *1912: The Year the World Discovered Antarctica* (London: Bodley Head/Pimlico, 2012).
Tyler-Lewis, Kelly, *The Lost Men* (London: Bloomsbury, 2006).
Walker, Caroline, *MacDonald Gill: Charting a Life* (London: Unicorn, 2020).
Weddell, James, *A Voyage Towards the South Pole … 1822–1824* (London: Longman et al., 1827).
Weems, John Edward, *Race for the North Pole* (London: William Heinemann, 1961).
Wheeler, Sara, *Cherry: A Life of Apsley Cherry-Garrard* (London: Vintage, 2001).
— *Terra Incognita: Travels in Antarctica* (London: Vintage, 1996).
— *The Magnetic North: Travels in the Arctic* (London: Vintage, 2010).
Wild, Frank, & Macklin, Alexander, *Shackleton's Last Voyage* (London: Cassell & Co., 1923).
Williams, Glyn, *The Quest for the Northwest Passage* (London: The Folio Society Ltd, 2007).
Williams, Isobel, *With Scott in the Antarctic: Edward Wilson* (Stroud: The History Press, 2008).
Williams, Isobel, & Dudeney, John, *William Speirs Bruce: Forgotten Polar Hero* (Stroud: Amberley Publishing, 2018).
Wilson, David M, *Nimrod Illustrated* (Cheltenham: Reardon Publishing, 2009).
— *The Lost Photographs of Captain Scott* (London: Little, Brown, 2011).
Wilson, David M., & Elder, David B., *Cheltenham in Antarctica: The Life of Edward Wilson* (Cheltenham: Reardon Publishing, 2000).
Wilson, David M., & Wilson, C.J., *Edward Wilson's Antarctic Notebooks* (Cheltenham: Reardon Publishing, 2000).
Wilson, Edward (edited by Harry King), *South Pole Odyssey: Selections from the Antarctic Diaries of Edward Wilson* (Cambridge/Poole: Scott Polar Research Institute/Blandford Press, 1982).
Worsley, Frank, *Shackleton's Boat Journey* (London: Pimlico, 1999).
Yelverton, David E., *Antarctica Unveiled* (Boulder, Colorado: University Press of Colorado, 2000).
— *Quest for a Phantom Strait: The Saga of the Antarctic Peninsula Expeditions 1897–1905* (Guildford: Polar Publishing Ltd, 2004).
Young, Louisa, *A Great Task of Happiness: The Life of Kathleen Scott* (London: Hydraulic Press, 2012).

Academic Papers, Journal and Other Articles

Alp, W.J., *Polar Record* papers (Cambridge University Press), including 'Dogs of the British Antarctic Expedition 1910–13' (June 2019).
Attwood, Philip, 'Kathleen Scott: The Sculptor as Medallist', *British Numismatic Society Journal*, Vol. 60, 1990–91 (online).

Basberg, Bjørn, & Headland, Robert, 'The 19th Century Antarctic Sealing Industry: Sources, Date and Economic Significance', Norwegian School of Economics and Business Administration (online, Norwegian Open Research Archives).

Bean, Kendra, 'Hidden Treasures of Our Collection: Herbert Ponting's Cine Camera', Science & Media Museum online blog, 23 January 2019.

Cook, Andrew, 'Harry Goodsir, Lost Naturalist of the Franklin Arctic Expedition', *New Orkney Antiquarian Journal*, Vol. 7 (John Rae 200 Conference proceedings, Kirkwall, Orkney Heritage Society, 2014).

Debenham, Frank, 'Retrospect: The Scott Polar Research Institute, 1920–45', *Polar Record*, Vol. 4, January 1945, pp. 223–35.

Dudeney, J.R., Sheil, J., & Walton, D.W.H., 'The British Government, Ernest Shackleton, and the Rescue of the Imperial Trans-Antarctic Expedition', *Polar Record*, Vol. 52:4, pp. 380–92 (2016).

Finnegan, Diarmid, 'Crozier's Penguin: an object history of maritime and museum scientist', *Endeavour*, Vol. 42 (1), pp. 42–47, Queen's University, Belfast.

Firth, P.G., Benavidez, O.J., & Fiechtner, L. 'The Signs and Symptoms of Ernest Shackleton', *Journal of Medical Biography* (2021).

Gould, William J., 'HMS Challenger and SMS Gazelle – their 19th century voyages compared', History of Geo- and Space Sciences Discussions, 2022 (online).

Guly, Henry, 'The understanding of scurvy during the heroic age of Antarctic exploration', *Polar Record*, Cambridge University Press (pub. online 30 September 2011)

Hood, Stephanie, 'Science, Objectivity and Photography in the Nineteenth Century: Photographs from the Voyage of HMS Challenger 1872–76', Royal Museums Greenwich blog (based on 2013 PhD).

Hyde, Ralph, et al. (inc. Russell Potter), 'Dictionary of Panoramists of the English-Speaking World' (online).

Jones, A.G.E., 'Sir James Clark Ross and the Voyage of the Enterprise and Investigator, 1848–49', *Geographical Journal*, June 1971, Vol. 137, No. 2, pp. 165–79, London (via JSTOR).

Kaufman, Matthew, 'Harry Goodsir and the last Franklin Expedition', *Journal of Medical Biography* (London, 2004), 12, pp. 82–89.

Lummis, Geraldine, 'Imaging Sir Joseph Kinsey (1852–1936): a man of many parts' (University of Canterbury thesis, 2019).

Markham, Clements, 'Rescue of the Swedish Antarctic Expedition', translation of Irízar's report of 21 December 1903, *Geographical Journal*, May 1904, Vol. 23, pp. 580–96 (London).

Mawson, Douglas, 'The B.A.N.Z. Antarctic Research Expedition, 1921–31', *Geographical Journal*, August 1932, Vol. 80, No. 2.

Montalbán, Christina, 'Shackleton: The Last Voyage', *Antarctic Affairs*, Vol. 2, pp. 37–55 (2015).

Plunkett, John, 'Moving Panoramas *c.* 1800 to 1840', *Interdisciplinary Studies in the Long Nineteenth Century*, 17 (2013), online.

Potter, Russell A., 'Icebergs at Vauxhall', *Victorian Review*, Vol. 36, No. 2, pp. 27–31 (The John Hopkins University Press, 2010).

Rosove, Michael, 'Who discovered the emperor penguin?', *Polar Record*, Vol. 54 (2018), pp. 43–52.

Rubin, Morton, 'Thaddeus Bellingshausen's Scientific Programme in the Southern Ocean', *Polar Record*, Vol. 21 (1982), pp. 215–29.

Savours, Ann & McConnell, Anita, 'The History of the Rossbank Observatory, Tasmania', *Annals of Science*, 39 (1982), pp. 527–64.

Sharpe, Tom, 'On the rocks on Elephant Island', *James Caird Society Journal*, No. 7 (2014), pp. 21–32.

Stein, Glenn M., 'Photography Comes to the Polar Regions – Almost', *Antarctic*, New Zealand Antarctic Society, Vol. 26, No. 1 (2007–13).

— 'The Arctic Medal 1818–55 to Members of the Antarctic Expedition of 1839–43', *Orders and Medals Research Society Journal (OMRSJ)*, September 2020.

— 'The Challenger Medal Roll (1895)', *OMRSJ*, December 2006/January 2007.

Stenton, Douglas R., 'Finding the dead: bodies, bones and burials from the 1845 Franklin northwest passage expedition', Cambridge University Press, 2018 (online).

Tomlinson, Barbara, 'Chivalry at the Poles: British Sledge Flags', *The XIX International Congress of Vexillology Proceedings* (York, 23–27 July 2001) pp. 215–22 (pub. Flag Institute, 2009 and online).

Yelverton, David, 'The Riddle of the Antarctic Peninsula' (Nordenskjöld expedition), *Antarctic*, New Zealand Antarctic Society, Vol. 15, No. 4, Vols 1 and 2 (all 1998).

Background reading also includes articles from *James Caird Society Journal* (James Caird Society); *Nimrod: The Journal of the Ernest Shackleton Autumn School* (Athy, Ireland); and *Polar Record* (Cambridge University Press).

Index

This index, which covers the main text, complements the Summary Timeline (Appendix B) by providing page references for individuals, ships, places and generic subjects in which readers might be particularly interested. Some individuals or other indexed subjects are world famous, others are less well known but merit inclusion due to, for example, pioneering roles or involvement in multiple expeditions. Omission from the index does not, however, imply a lack of contribution to polar history. Geographical features are listed (subject to caveats mentioned in Appendix A) using names which allow them to be found in Appendix C or other readily available maps. Vessels are indexed alphabetically by name, without 'HMS' or similar designations; references to vessels include those to expeditions named for them, e.g. Scott's *Terra Nova* expedition, Shackleton's *Endurance* expedition. Where surnames are identical or very close, family relationships or additional information is provided. Pages containing illustrations of indexed subjects are indicated by italicised numbers, unless those pages fall within a group of pages or on pages also containing text on the same subject.

Adams, Jameson 157, 159–60
American Geographical Society 193, 233
Amundsen, Roald 122–5, 151, 152–6, 173, 178, 179–82, 187–90, 191–3, 197–8, 201, 204, 217, 223, 227, 228, 230–3, 236, 238, 242, 263
Andresen, Adolphus and Wilhelmina 'Mina' 163–6
Antarctic Heritage Trust (New Zealand) 126, 255, 258
Armitage, Albert 118–19, 131
Arner Boyd, Louise 233
Atkinson, Dr Edward 197–8, 256–7
Aurora 181, 183, 185–6, 202, 204, 236, 256–7
Aurora Australis 15, 38, 63, 126, 138, 177

Austin, Horatio 84, 87, 103
aviators – *see* Amundsen, Roald; Campbell, Stuart; Carr, Roderick; Douglas, Eric; Gran, Tryggve; Nobile, Umberto; Wilkins, Hubert

Back, George 30, 51, 95
Baffin Bay 20, 26, 28–9, 34–5, 56, 72–3, 75, 78–9, 103, 152
Banks, Joseph 11, 13, 18, 21, 25–6, 34–5, 37–8
Barrow, [later Sir] John 25, 28–31, 35, 37, 69, 83, 84, 263
Barrow, John (son of above) 84, 110
Barrow Strait 43–5, 47, 50, 69, 74, 76, 78, 83–4, 87, 106

Bay of Whales 181–2, 232, 238
Beardmore, William 157, 160
Beardmore Glacier (previously 'Great Glacier') 159, 204
Beechey, Frederick 27–8, 30, 84
Beechey Island 84–7, 90, 152, 251
Belgica 122–5
Bellingshausen, Fabian Gottlieb von 37–8
Bering Strait 20, 31, 34, 69, 74–5, 78–9, 81, 106, 155, 173
Bernacchi, Louis 126, 128–30, 131–5, 138, 210, 248
Booth, Felix 47, 50, 84
Boothia 48–50, 55, 77, 79–80, 83, 90, 154
Borchgrevink, Carsten 126, 128–30, 174
Bowers, Emily (mother of Henry) 196, 199
Bowers, Henry 'Birdie' 178, 187, 189, 195–9, 207, 209, 212, 230, 256–7
Bowers, Mary (sister of Henry, later Lady Maxwell) 199, 248, 250
Bransfield, Edward 37
Browne, William H. 75–8
Bruce, William Speirs 119, 141, 143–5, 150, 173, 201, 204, 206, 212, 216
Buchan, David 25–6, 27–8, 45
Byrd, Richard 217, 238, 242

Caird, Sir James 205–6, 216; see also *James Caird* (vessel)
Campbell, Stuart 234–7
Cape Adare 60, 126–30, 132, 179, 181, 185–6
Cape Crozier 63, 68, 132–4, 138–9, 178, 189, 207, 209, 257–8
Cape Evans 177, 187, 195, 197, 255–7
Cape Royds 159, 177, 255
Cape Town (inc. Cape of Good Hope) 13, 20, 57, 64, 97, 218–19, 221, 234
Carr, Roderick 'Roddy' 218–22
Central News Agency/CNA 187, 189, 198
Challenger 95, 96–9, 121
Charcot, Jean-Baptiste 147, 150, 162–6, 204, 232–3
Cheetham, Alf 204, 216, 221
Cherry-Garrard, Apsley 178, 209–10, 256–7

Clayton, Phillip 'Tubby' 260–2
Clerke, Charles 20
Colbeck, William 126–7, 134–5, 234
Collinson, Richard 78, 84
Conan Doyle, Arthur (inc. as Arthur Doyle) 110–11, 115
Cook, Elizabeth (wife of James) 18–22
Cook, Frederick 122, 124–5, 165, 167–70, 173, 193, 230, 232
Cook, James 11–12, 13–17, 18, 20–2, 23, 34, 37–8, 39–40, 68, 96, 124
Cracroft, Sophia/Sophy 63, 68, 70, 90–1, 250
Crean, Tom 140, 189, 197, 204, 214, 263
Crozier, Francis 45, 56, 57–60, 61, 63–4, 65–8, 69, 70, 72–4, 76–8, 88–91, 94, 251

David, Edgeworth 160, 179–82
Davies, Frankie 176, 255
Davis, John (*Terror* 1839–43) 59, 61–2
Davis, John K. (*Aurora*) 183, 186, 202
de Gerlache, Adrien 122–5, 126, 202
Debenham, Frank 197, 239, 241–3, 246–8, 250
Deception Island 150, 162–6
Discovery (Cook) 20
Discovery (Nares) 100–2
Discovery (Scott et al.) 68–9, 131–5, 136, 138–40, 144–5, 157, 182, 189, 204, 207, 225, 234–8, 241–3, 248–9, 255, 260; see also Mawson, Douglas; Scott, Robert; Watkins, Gino
Docker, Dudley (inc. eponymous boat) 204, 213–14
Douglas, Eric 234–8
Drygalski, Erich von, 141, 144
d'Urville, Jules Dumont 59–60, 63–4

Elephant Island 212–16, 220
Ellsworth, Lincoln 228, 230, 238
Enderby Land (inc. Enderby Brothers, John Biscoe) 42, 236, 238
Endurance 202–6, 212–13, 218, 220–1, 245, 248–9, 256; see also Shackleton, Ernest
Enterprise 74–8, 84, 106

Erebus 57, 60, 61, 63–4, 65–8, 69, 70–3, 82, 86–7, 91, 94, 140, 249, 251–4; *see also* Franklin, John; Ross, James Clark
Evans, Edgar (*Discovery, Terra Nova*) 146, 178, 187, 195, 197–8
Evans, Edward ('Teddy', *Terra Nova*) 135, 189, 195, 197–8, 209
Evans, Hugh (*Southern Cross*) 126, 128

Falkland Islands/the Falklands (inc. Stanley) 39–40, 63–4, 141, 206, 262
Farthest North 23, 46, 100, 102, 119, 170, 206
Farthest South 39–40, 60, 63, 124, 129, 157, 159, 161, 165–6, 180, 182
Filchner, Wilhelm 173, 189, 201, 204, 206
films and filming 128, 145, 161, 174–8, 180, 182, 204, 242, 260; *see also* Bernacchi, Louis; Gaumont; Hurley, Frank; Ponting, Herbert; Wilkins, Hubert
Fitzjames, James 69, 91, 94, 251
Forbes, Edward 65, 71–3
Forster, Johann (father) and Georg (son) 13–14, 65, 68
Fram 116, 118–19, 133, 151, 173, 181, 193, 201, 230, 249; *see also* Amundsen, Roald; Nansen, Fridtjof
Franklin, Eleanor (John Franklin's daughter by first wife), later Gell 63, 68, 75, 88–94
Franklin, Jane (Franklin's second wife) 56, 57, 59–60, 63, 68, 70–2, 74–5, 78, 84, 88, 90–2, 95, 104, 110, 250
Franklin, John 25–6, 27–8, 30, 49, 57–60, 63, 67–8, 69, 70–3, 74–8, 80–3, 84–7, 88–91, 83, 94, 95, 102–3, 104–8, 115, 121, 128, 152–4, 201, 209, 230, 248, 250, 251, 254, 263
Franz Josef Land 111–12, 116, 118–20
Fuchs, Vivian 258, 259
Fury (inc. Fury Beach) 45, 47–51

Gaumont 161, 174–5, 177–8, 217
Gell, Eleanor (née Franklin) – *see* Franklin, Eleanor
Gell, Philip (husband of Eleanor) 88–93

Gill, MacDonald 248–9
Gjøa (inc. Gjoa Haven) 152–6, 230
Goodsir, Harry 70–3, 85–7, 94
Goodsir, Robert (brother of Harry) 85–7
Gore, Graham (with Franklin) 91, 94, 251
Gore, John (grandfather of Graham, with Cook) 13, 20
Graham Land 42, 46–7, 150, 243
Gran, Tryggve 217, 232
Gray, David and John 110–11, 114
Grinnell, Henry 84, 105–6

Henson, Matthew 167–71
Hobart, Tasmania 57–60, 63, 68, 69, 88, 129–30, 181, 183, 187, 236, 238
Hodges, William (*Resolution* artist) 13–17
Hodgeson, Frederick (*Challenger* photographer) 96–7, 99
Hodgson, Thomas (*Discovery* scientist) 131, 134–5
Hooker, Joseph 65–6, 68, 140
Hudson's Bay 20, 30–1, 44, 79
Hudson's Bay Company/HBC 30, 34, 72, 79–83, 84, 90, 251
Hulott, Cornelius 104–9
Hurley, Frank 184–6, 205, 215, 217, 234–8, 261–2
Hussey, Leonard 218, 220–1
Hutchison, Isobel Wylie 248, 250

icebergs and 'ice islands (notable encounters) 13–17, 63–4, 73, 96–9, 138, 145, 177, 189
Inuit 30, 34, 48–50, 55, 76, 79–82, 84, 86, 90, 103, 103, 107, 151, 154, 167, 170, 241, 244, 245, 251, 253
Investigator 74–8, 84, 104–7, 109
Irízar, Julián 146–51, 213
Irving-Bell, Dorothy *see* Russell Gregg, Dorothy

Jackson, Frederick 116, 117, 118–20, 131
James Caird 213–14, 216, 249
Jameson, Professor Robert 25, 32, 34, 41
Johansen, Hjalmar 119, 193

Kennet, Lady Kathleen – *see* Scott, Kathleen
Kerr, Alexander (*Endurance, Quest*) 218, 221
Kerr, Gilbert (*Scotia*) 143–5
King Edward [VII] Land 133, 138, 161, 173, 180, 182
King William Land/Island 49, 90–1, 154, 251, 253
Kinsey, Joseph 132, 175, 177, 181, 187–90, 198, 209, *210*
Koettlitz, Reginald 118, 131, 134–5

Lancaster Sound 28–30, 34, 36, 43, 45, 47, 51, 69, 74, 76, 78, 152
Larsen, Carl 146, 148–50
Lashly, William 189
Lay, Jesse 99
Lefroy, [John] Henry (cousin of Jessie Lefroy) 57, 79–80, *81*
Lefroy, Jessie and other family members 57, 70, 91, 248, 250
Leigh Smith, Benjamin 110–15, 116, 118–20
Levick, George Murray 148, 258
Lowe, George 257–8
Lyall, David 65–6, 68

Macklin, Alexander 218–20, 259
Markham, Albert (cousin of Clements) 102–3, 131
Markham, Clements 87, 95, 103, 115, 118–19, 121, 130, 131–2, 135, 161, 199, 207, *210*
Marr, James 'Scout' 215, 225–6, 234, 248, 259–61
Marshall, Eric 157–60
Mawson, Douglas 160, 180–1, 183–6, 189, 202, 204–5, 217, 234, 236–8
McClintock, F. Leopold 75, 77, 91–2, 100, 102–3, 110, 121, 128, 131, 153
McClure, Robert 75, 78, 84, 104, 106–9
McCormick, Robert 65–6, 68
McDonald, Dr Alexander 85–6
McIlroy, James 216, 218–19

McMurdo Sound 59–60 (unnamed), 129, 132–3, 135–6, 138, 157, 159–61, 176, 187, 206, 255–7
McNish, Harry (carpenter, aka 'Chippy) 213, 216
Meares, Cecil 174, 188, 209, 256–7
Morning 132, 134–5, 138, 140, 234
Mossman, Robert 144–5, 150
Mount Erebus 61, 63, 129, 133–4, 160, 176
Mount Terror 61, 129, 133–4, 139, 176
Murray, John 96, 99, 121, 145

Nansen, Fridtjof 116, *117*, 118–20, 131, 151, 152, 155–6, 173, 178, 193, 201, 204, 209, 233, 263
Nares, George 96–7, 99, 100–3, 110, 115, 131
National Geographic Society (US) 171, 191, 232
Newbold, Caleb 96–7, 99
Newman, Arthur S. (inc. eponymous cameras) 128, 174, *176*
Nimrod 159–61, 177, 180, 187, 191, 193, 202, 204, 212, 223, 225, 246, 255
Nobile, Umberto 228, 230, 232
Nordenskjöld, Otto 141, 144, 146–50, 201, 204, 206, 213
North Magnetic Pole 26, 47, 49–50, 53, 56, 63, 77, 151, 153–4, 156, 171
North Pole (geographic) 26, 27, 34, 37, 45–6, 95, 100, 102–3, 110, 116, 118–19, 151, 165, 167–71, 173, 178, 189, 191, 193, 201, 217, 227, 230, 232
Northwest Passage 11, 20, 26, 30, 34, 37, 43, 45–6, 47, 52, 69, 73, 74, 79, 84, 95, 106, 109, 151, 152–3, 156, 191, 228, 230

Oates, Lawrence 187, 195, 197–8, 230, 233, 256–7
Orkney (inc. Stromness) 72, 79–80, 82–3, 92

panoramas (inc. owners and painters) 27–31, 55, 78
Palliser, Hugh 11, 18, 20

Parry, (William) Edward 26, 28–31, 34, 37, 43–6, 47, 49, 52, 56, 60–1, 63, 67, 69, 74, 80, 84, 106, 108, 263
Paulet Island 149–50, 213
Peary, Josephine (wife of Robert) 167, 179
Peary, Robert 151, 165, 167–71, 173, 191–3, 198, 204, 209
penguin species (identification, study of), 14, 17, 63–4, 65–8, 99, 128, 134, 136, 138–40, 144–5, 166, 176–7, 206, 213–14, 226, 230, 236–7, 246, 255, 257
penguins (as food, anti-scorbutic) 124, 126, 143, 214, 230; see also scurvy
Pennell, Harry 187–9, 195, 197–9
Penny, William 85, 87
Phipps, Constantine 18–19, 45–6
photographers (on expeditions) see Bernacchi, Louis; Cook, Frederick; Hodgeson, Frederick; Hurley, Frank; Lay, Jesse; Newbold, Caleb; Ponting, Herbert; Skelton, Reginald
Pirie, James 141–5
Ponting, Herbert 174–8, 188, 198, 209, 216, 227, 248, 255, 257–8, 261–2
Priestley, Raymond 212, 239, 246, 248
Prince Regent Inlet 43, 45, 47, 50, 69, 74, 76, 78, 84

Quest 218–22, 223–6, 232, 234, 241–3, 245, 259–62

Rae, Dr John 72, 79–83, 90–2, 95, 152
Resolute 84, 96, 104–6, 108–9
Resolution (James Cook) 13–16, 18, 20–1, 23
Resolution (Scoresbys) 23–5
Richardson, Dr John 30, 73, 74, 80–1, 88
Ross, James Clark (nephew of John) 30, 45, 47–51, 52–6, 57–60, 61–4, 67–8, 69, 74–8, 80, 84, 88, 96, 99, 106, 124, 129–30, 133, 144, 154
Ross, John 26, 28–30, 34–6, 43, 47–51, 52–6, 74, 80, 84
Rowett, John Quiller 218, 221, 226, 260
Royal Geographical Society/RGS (inc. RGS medals) 42, 52, 67, 83, 92, 95, 102–3, 110, 112, 114–15, 116, 121, 129–30, 131, 145, 156, 161, 171, 190, 192–3, 202, 204, 209, 239–44, 258
Royal Scottish Geographical Society 145, 239, 243
Royal Society 11–13, 18, 20–2, 25, 28, 44, 56, 67, 95, 102, 130, 204
Royal Society of Edinburgh 34–6, 41
Royds, Charles 131, 133–4, 138–9
Rudmose Brown, Robert 141–2, 145
Russell Gregg, Dorothy (later Irving-Bell) 223–7
Rymill, John 244, 248

Sabine, Edward 28–30, 50, 56, 57, 60, 64, 67
Scoresby, William (of Whitby) 23–6, 34, 46
Scoresby, William (son of William) 23–6, 32–6, 37, 41, 46
Scott, Kathleen (wife of Robert, later Kennet/Hilton-Young) 192, 198–9, 207–11, 248, 250
Scott, Robert 68, 131–5, 136, 138–40, 144, 157, 159, 161, 162, 166, 173, 174–5, 177–8, 179–82, 187–90, 192–3, 195, 197–8, 201, 204, 207, 209, 230, 246, 248, 255–6, 258, 263
Scott Polar Research Institute/SPRI 57, 70–1, 199, 244, 246–50
scurvy and prevention of 12, 18, 20, 34, 49–51, 77, 100, 102–3, 107, 115, 124, 126, 134–5, 143, 145, 189, 214, 248
seals and sealing 17, 23, 26, 34, 37, 39–42, 55, 63, 71, 122–5, 126, 136, 143, 162, 176, 196, 214, 222, 236, 244, 259; *see also* South Orkneys; South Shetlands; Weddell, James
Service, Robert 223–7
Shackleton, Emily (wife of Ernest) 206, 220, 226–7
Shackleton, Ernest 131, 133–5, 146, 157–61, 165, 177, 180, 187, 191–3, 198, 201, 202, 204–6, 212–14, 216, 217–22,

223–7, 234, 241, 248, 255–6, 259–60, 262, 263; *see also Discovery*; *Nimrod*; *Endurance*; *Quest*
Shetland Islands (inc. Fair Isle) 26, 141–3
Shirase, Nobu/Japanese Antarctic Expedition 179–82, 189, 232
Skelton, Reginald 131, 134–5, 138–40, *210*
Smith, William 37, 39
Snow Hill Island 64, 146–50, 213
Sobral, José 146–8, 150
South Georgia 17, 40, 150, 157–8, 206, 213–14, 216, 219–21, 262
South Magnetic Pole 58–60, 63, 122, 124, 129, 138, 157, 160, 185–6
South Orkney Islands 39–41, 143
South Pole (geographic) 16, 119, 157, 161, 173, 178, 182, 187, 189, 192–3, 195, 197–9, 201, 206, 207–9, 227, 230, 246
South Shetland Islands 37, 39–40, 64, 162, 212
Southern Cross 126–30, 131, 133, 134, 136, 174, 248
Speirs Bruce, William – *see* Bruce, William Speirs
Spitsbergen 20, 26, 27–8, 30, 32, 45–6, 110–11, 201, 216, 232
Stancomb-Wills, Janet (inc. eponymous boat) 204, 213–14
Stefansson, Vilhjalmur *243*

Tasmania, see Hobart
Terra Nova 68, 140, 166, 174–7, 187–9, 193, 195, 197, 199, 202, 212, 217, 222, 227, 246, 248, *249*, 255–8; *see also* Scott, Robert
Terror 57, 59–60, 61, 63–4, 65, 69, 70, 72, 82, 85–6, 91, 94, *249*, 251, 253–4; *see also* Crozier, Francis; Franklin, John; Ross, James Clark

Watkins, Henry 'Gino' 239–44, 248
Weddell, James 39–42, 60
Weddell Sea 42, 64, 143–4, 173, 201, 204, 206, 212, 214, 216, 241
Weddell seal/sea leopard 39–41, 122–5, 196
whales and whaling 23–6, 32, 35, 37, 51, 56, 75, 84–5, 87, 105, 110, 114–15, 118, 122, 131–2, 140, 146, 150–1, 154, 162–6, 176, 183, 206, 237, 251; *see also* Deception Island; Gray, David and John; Larsen, Carl; Scoresby, William (father and son), South Georgia; Weddell, James
Whitby, Yorkshire 11, 13–14, 20, 23–6, 132–5
Wild, Frank 133, 157, 159–60, 183, 186, 204–5, 214, 216, 218, 220–1, 225–7, 259–61
Wilkes, Charles 59–60, 63, 237 (Wilkes Land)
Wilkins, Hubert 166, 217, 218, 221, 232, 234, 259, 261
Wilson, Edward 68, 131, 134–5, 136–40, 145, 177–8, 187, 189, 195, 197–9, 207–11, 230, 248, 257, 263
Wilson, Oriana (née Souper, wife of Edward Wilson) 187–90, 198–9, 209–11, 248, 250
Wilton, David 118, 141–2
Wordie, James 212–14, 216, 239, 241, 243, 248
Worsley, Frank 205, 213–14, 218, 225

By the same author …

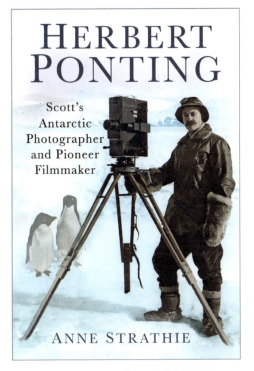

PAPERBACK: 978 0 7509 7901 6
EBOOK: 978 0 7509 9705 8

'Working together with paint box and camera, Edward Wilson and Ponting created the aesthetic to define a new continent: Antarctica! Here, at last, is a major new biography of one of our greatest photographers, Herbert Ponting.'

Dr David Wilson

By the same author …

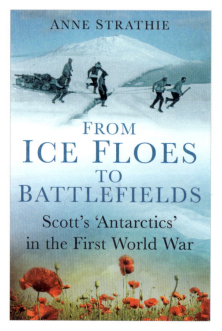

'[Strathie's] book is well written, excellently illustrated … and recommended to all those who want to know more about Scott's Antarctic survivors and their Great War Deeds.'

Gallipoli Magazine

EBOOK: 978 0 7509 6578 1

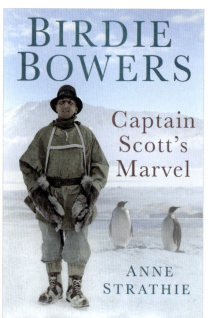

Henry 'Birdie' Bowers was a stocky bundle of energy, knowledgeable, indefatigable and the ultimate team player. In Scott's words, he was 'a marvel'.

PAPERBACK: 978 0 7524 9444 9
EBOOK: 978 0 7524 7871 5

The destination for history
www.thehistorypress.co.uk